DNA

FOR BEGINNERS

Writers and Readers

DNA... three capital letters of which perhaps many may have heard, but not so many will have understood. This book is about the discovery of the significance of DNA, beginning in the mid-nineteenth century. We take a close look at the mechanisms believed to be important in its functioning, recognizing that much remains shrouded in mystery. And we examine the impact of DNA research on society, and some of the most recent findings.

Most DNA is in the form of a helical molecule. Some DNAs are very long, and some are small, but all DNA is so tiny that huge amounts of it can be found in the cells of living things. DNA holds the codes for an enormous variety of genes. And genes are the pieces of information which we all have inside us, enabling us to function and reproduce.

To get some idea of the great significance of DNA in the history and future of life on this planet, read on...

In December of 1949, almost four years before James Watson and Francis Crick published their model of DNA, launching the revolution in modern biology, the mathematician and designer of the computer, John von Neumann, gave a lecture explaining how a machine could reproduce. All it needs, he said, is a description of itself.

A machine with a magnetic core could not reproduce the magnetic core by making a mould. However, if it had a

description reading, "magnetic core: electric wire tightly wound around metal bar five hundred times, etc." and it had the necessary raw material, it could easily follow the description and build the magnetic core.

The machine's offspring could reproduce as well, if the machine made a **copy** of the **description of itself** and inserted that copy into every new machine. Given the necessary raw materials, the machines could go on making copies of themselves.

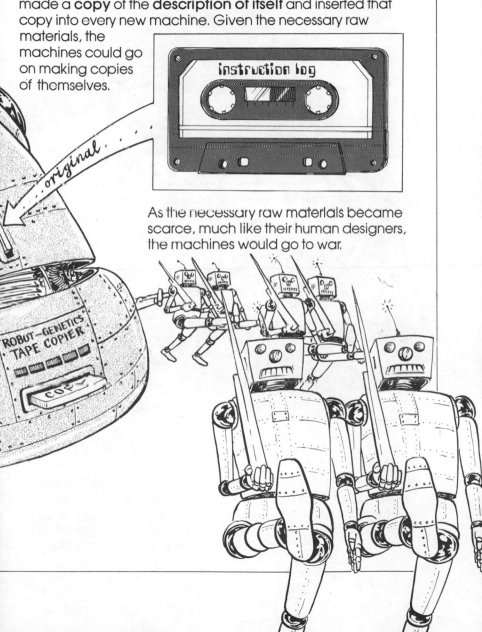

As the necessary raw materials became scarce, much like their human designers, the machines would go to war.

The **description** in von Neumann's machine is analogous to the DNA found in living things. Like the description of the machine, DNA contains the coded description of the organism and is responsible for its capacity to reproduce.

Living things, unlike von Neumann's machine, do not usually make **exact** copies of themselves. For then, there would be no evolution and life as we know it would not exist. Living things make **variant** copies of their parent organism or organisms. They can do this because of DNA.

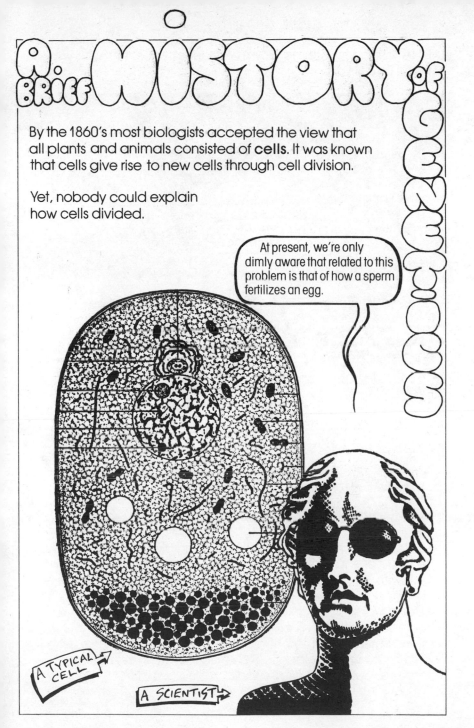

A BRIEF HISTORY OF GENETICS

By the 1860's most biologists accepted the view that all plants and animals consisted of **cells**. It was known that cells give rise to new cells through cell division.

Yet, nobody could explain how cells divided.

> At present, we're only dimly aware that related to this problem is that of how a sperm fertilizes an egg.

A TYPICAL CELL

A SCIENTIST

DNA was discovered in 1869 by Frederick Miescher, who was then 25 years old.

Miescher was the son of a well known physician in Basel. In 1869 he had gone to Tubingen to study the chemistry of white blood cells with the biochemist F. Hoppe-Seyler. He used pus obtained from postoperative bandages, as a source of the cells. When he added weak hydrochloric acid to the pus he obtained pure nuclei. If he added alkali and then acid to the nuclei a grey precipitate was formed. The precipitate was unlike any of the known organic substances. Since it came from the nucleus, Miescher called it **nuclein**. Today it is called DNA.

Shortly after Miescher's discovery, new staining techniques were developed which revealed band-like structures in the nucleus of the cell that stained very darkly. In 1879 Walter Flemming introduced the term **chromatin** (Chroma: Greek for color) to describe the intensely stained material in the nucleus. In 1881 E. Zacharia found that chromatin reacted to acid and alkali in the same way as Meischer's nuclein. He concluded that nuclein and chromatin were one and the same.

TUBINGEN CASTLE
BIOCHEMICAL LAB.

The chromatin material observed in the 1880's was what are today called chromosomes, the carriers of genes that are the basis of heredity. What is most remarkable, is that some scientists studying fertilization made the connection between chromatin (chromosomes) and heredity already in the 1870's. Using the light microscope, Hermann Fol in Switzerland and Oskar Hertwig in Berlin independently observed that the sperm **penetrates** the egg and that the **nuclei** of the sperm and the egg **fuse**. And Edouard Van Beneden, studying the threadworm Ascaris (a parasite of horses) noted that the sperm contributed the same number of chromosomes as the egg to the developing embryo. He also discovered **meiosis**, the halving of the number of chromosomes in the germ cells (the egg and the sperm). It was Flemming who observed cells dividing and saw chromosomes replicating. He concluded that chromosomes were a source of continuity from one generation to the next.

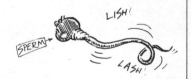

So by the 1890's scientists had come to have a clear idea of the nature of fertilization, and were even declaring that DNA (Meischer's nuclein) was the basis of heredity.

Modern genetics begins with Gregor Mendel's famous
experiments with garden peas in the 1860's. Mendel had chosen
peas that had certain pure traits which always breed true. He
had plants that produced yellow seeds, and others that
produced green seeds. When he cross-bred these types, all the
progeny were yellow-seed bearing plants. Mendel called the
yellow trait DOMINANT, and the green RECESSIVE. He argued
that the progeny of these first generation crosses had each
received an equal genetic contribution from each parent, but
only the dominant yellow trait was manifested.

When he crossed these first generation hybrids with each other he found that 75% of the progeny were yellow and 25% green, confirming his supposition that the green "gene" (he didn't use this term) had been there all the time. Mendel concluded that green and yellow were discrete genetic units which assorted independently according to the laws of chance.

If chance governed the laws of inheritance of genetic information, where did the information needed to form a complicated biological organism reside? One answer, the so-called Vitalist School of Thought, argued that living organisms were shaped from outside by the hand of God. The Vitalists believed in EPIGENESIS, that the embryo developed from a simple unformed egg, gradually becoming a complex organism.

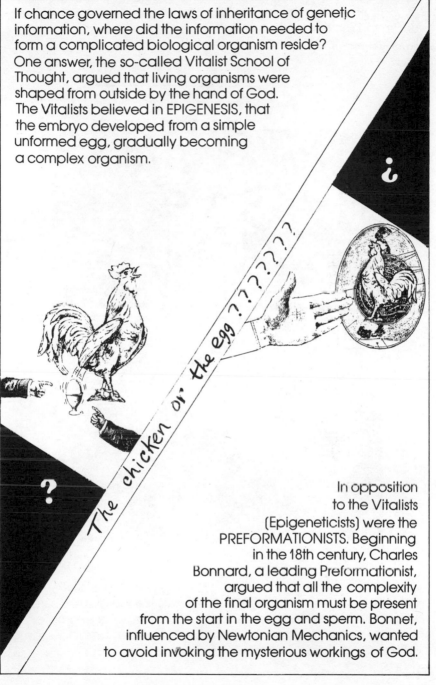

The chicken or the egg ? ? ? ? ? ? ? ?

In opposition to the Vitalists (Epigeneticists) were the PREFORMATIONISTS. Beginning in the 18th century, Charles Bonnard, a leading Preformationist, argued that all the complexity of the final organism must be present from the start in the egg and sperm. Bonnet, influenced by Newtonian Mechanics, wanted to avoid invoking the mysterious workings of God.

By the 19th century, biologists could see the embryos develop under the microscope. Ernst Haeckel summarized a third hypothesis with a familiar phrase …

Ontogeny recapitulates phylogeny.

… This means the early development of the embryo seems to repeat the **adult** stages of lower life forms from which it has descended. For example, at an early stage in its development, the human embryo has gills like a fish. As stated, Haeckel's law is no longer accepted.

Nineteenth century scientists also wondered about the effect of environment on development. Mountain people have more red blood cells than their relatives living at sea level. And some people grow taller than others because of their diet. In 1809, Lamarck argued, among other things, that:

Changes brought about in an organism because of environment can be inherited. A tree that has been bent by the wind will give rise to bent trees.

This argument is usually summarized by the phrase: "acquired characteristics are inherited."

Modern genetics has disproved Lamarck. In Darwinian theory, **natural selection** – environmental pressures that affect an organism's chances of survival – operate on the innumerable variations among organisms.

WITHOUT VARIATION ALL INDIVIDUALS WOULD BE IDENTICAL AND THERE WOULD BE NO EVOLUTION.

The most obvious mechanism for creating genetic variation is sex. As with Mendel's peas, genetic traits are randomly assorted during sexual reproduction. The combinations, in a complex organism, are almost limitless.

When Darwin published **The Origin of Species**, the mechanism of fertilization of an egg by a sperm was not well understood. Darwin's explanation of fertilization (and the consequent variations) was derived from an old Greek theory: Pangenesis. Every organ and tissue secreted granules, called gemmules, which combined to make up the sex cells.

But Darwin's cousin, Francis Galton, transfused blood from rabbits of one color to those of another. The transfusions had no effect on the color of the offspring. Galton argued:

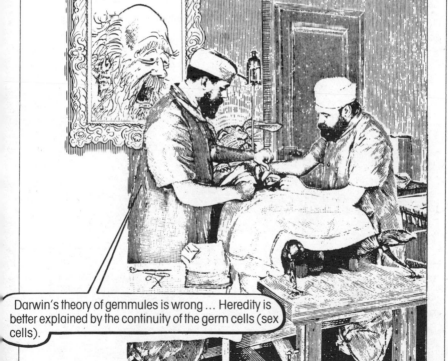

Darwin's theory of gemmules is wrong … Heredity is better explained by the continuity of the germ cells (sex cells).

This argument, subsequently developed by August Weismann asserted the continuity of the germ plasm.

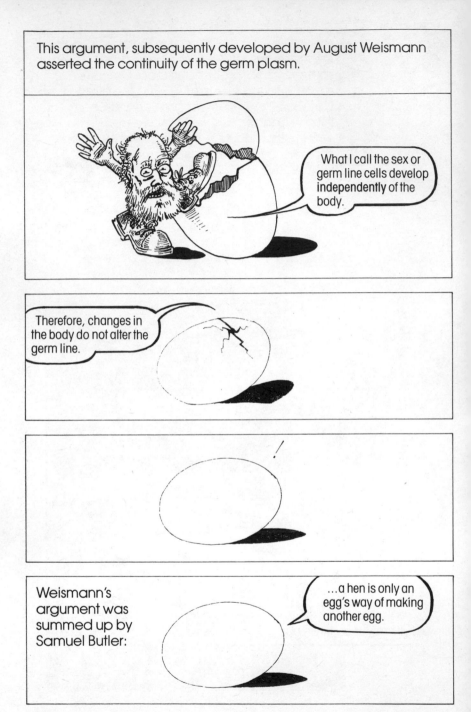

What I call the sex or germ line cells develop **independently** of the body.

Therefore, changes in the body do not alter the germ line.

Weismann's argument was summed up by Samuel Butler:

...a hen is only an egg's way of making another egg.

Modern terminology describes this new understanding:

1) THE GENE:
   at the time considered a hypothetical unit that was responsible for carrying genetic information from one generation to the next.

2) PHENOTYPE:
   the observed properties of a living thing; a consequence of the interaction of genetic makeup with the environment.

3) GENOTYPE:
   the genetic makeup of the organism, as opposed to its physical appearance.

Let us recapitulate what beliefs were held by the early decades of the 20th century:

1) Mendel showed that genetic traits were discrete units which assorted independently.

2) To explain development, Bonnet and the Preformationists had argued that complexity must be present in the egg and the sperm. The epigeneticists argued that the embryo developed from a simple to a complex form shaped by mystical forces.

3) Haeckel suggested that ontogeny recapitulates phylogeny.

4) Darwin's theory of evolution is based on the notion of **variations** (produced in part through sexual reproduction) that are selected by environmental forces **(natural selection)**.

5) Lamarck asserted inheritance of acquired characteristics.

A controversy over the mechanisms of evolution and development was raging. It could not begin to be resolved until scientists understood DNA.

At the turn of the century genetics was a young discipline. The early geneticist was little more than a statistician of inheritance. To speed experiments, geneticists turned to the fruit fly **drosophila**. The **drosophila** life cycle takes about 14 days, and the flies are easy to breed, cheap to grow, and genetically rather simple because they contain only four chromosome pairs.

The early drosophila geneticist, Thomas Hunt Morgan, observed violations of Mendel's Law of independent assortment of genes. Certain genes remained linked together in crosses more frequently than predicted by Mendel's statistics.

Morgan found four groups of linked genes in drosophila. The tendency of genes to remain together in offspring suggested that they shared a physical association, and were joined together on the same chromosome. Drosphila had four "linkage groups" because it had four chromosomes.

Morgan next observed a low frequency of "assortment" even for traits that were on the **same** chromosome. Morgan suspected that chromosomes could break and recombine, allowing genes on the same chromosome to reassort. Genes far apart on a chromosome would have a greater chance of a break occurring between them than genes situated close together. If so, reassortment frequency would be a measure of gene distance. Using this prediction to test his break and rejoin model, Morgan was able to make maps of genes on the **drosophila** chromosomes. Morgan's important finding was that genes fall in a defined linear order, and occupy specific positions in chromosomes.

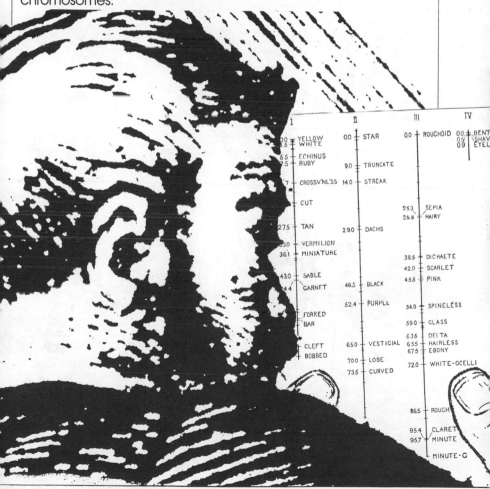

In the 1930's, scientists had little hope of "grinding a gene in a mortar, or distilling it in a retort". However, one property of a gene which could be analyzed was its ability to **mutate**. Hans J. Muller, the student of Morgan, increased the mutation rate of **drosophila** 15,000 fold over the natural rate by X-ray irradiation of the flies. Muller appreciated that mutations resulted from chemical reactions, or "sub-microscopic accidents" produced by the X-ray beam in the genetic material. Mutations never observed in nature, such as "splotched wing" and "sex-combless", as well as natural ones including "white eye", "miniature wing", and "forked bristles" were found.

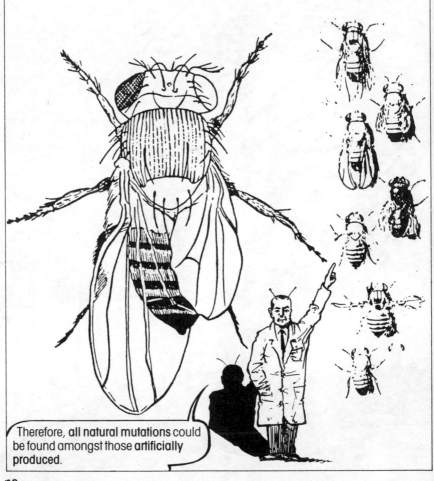

Therefore, **all natural mutations** could be found amongst those **artificially produced**.

Muller moved to Russia from the United States, where he disputed with the Soviet geneticist T.D. Lysenko.

I take the Lamarckian view that environmentally acquired characteristics are inherited.

No! After all, "white eye" and other natural mutations appeared in the "unnaturally" X-ray mutagenized flies. Lysenko's "environment" wasn't required to produce the mutations. Only the chemical changes produced by the X-rays were necessary. Mutations, either natural or lab-induced, provide the raw material for Darwinian selection.

Muller's experiments proved Lamarck and Lysenko erroneous.

Genetic damage can result from modern uses of radiation. I would also advocate eugenics, the regulation of reproductive behaviour to favor special physical, intellectual and moral qualities.

Although the dispute with Lysenko drove Muller from Russia he continued to speak out on the factors which shape the genetic constitution of society.

Puzzlement remained about how a gene actually operated. It would soon be shown, first by Sir Archibold Garrod in London, and later by the Americans George Beadle and E.L. Tatum, that a gene specified an enzyme.

An enzyme is a type of protein that catalyzes biological reactions. By specifying an enzyme, a gene enables a particular chemical reaction to take place in a cell.

I summarize this in my `one gene   one enzyme' hypothesis.

Ah, but now we must identify the chemical substances in chromosomes which store and transmit genetic traits: i.e. the genes.

Puzzled? So were they for a while. Ironically, the answers came, not from geneticists, but from the medical community.

Pneumonia is caused by the bacterium pneumococcus. Only certain strains of pneumococci are disease causing or virulent. The difference between virulent and non-virulent strains is a hereditary property of the strain, and a major question was the chemical basis for the biological specificity of virulence.

Two medical researchers studying pneumococcus provided the discoveries identifying DNA as the hereditary material. Both were shy, meticulous, and slight of build.

One, Fred Griffith, who worked at the Ministry of Health in London, was a taxonomist who devoted his career to developing reliable techniques for classifying pathogens (any organisms that cause disease).

The second, Oswald Avery, was the son of a mystical English pastor who immigrated first to Halifax, Nova Scotia, and later to New York. Avery worked at the Rockefeller Institute near the laboratory of the famous DNA chemist P.A.T. Levene.

Griffith, working in London, found that non-virulent strains formed **rough** colonies on agar plates. Virulent strains formed **smooth** colonies. The appearance of the colonies was enough to distinguish the two types. To assay the virulence of a colony, Griffith injected the bacteria into mice.

Griffith found that heat treatment of virulent bacteria killed them. They lost their ability to cause disease. But surprisingly, after injection of killed virulent (smooth) bacteria mixed with living non-virulent (rough) pneumocci, the mice died. Furthermore, the bacteria isolated from the diseased mice were the smooth virulent type, although the only living bacteria injected were rough non-virulent. Rough bacteria were converted to smooth by a non-living extract of the smooth bacteria. Most important, the change was a permanent, inheritable one. The gene which determined rough colonies versus smooth colonies had been altered.

Avery, excited by this discovery of imparting a hereditary change on pneumococci, sought to identify the component of the killed smooth pneumococcus which conferred the virulent type. The difference between the smooth and rough colonies, and the virulent versus the non-virulent strains, provided assays for the change.

NON-VIRULENT (ROUGH COLONIES)

VIRULENT (SMOOTH COLONIES)

HEAT-KILLED SMOOTH

Avery's lab adapted Griffith's mixed injection procedure so it worked under more defined laboratory conditions. Living rough pneumococci were mixed with heat killed

SERUM (ANTI-ROUGH)

smooth pneumococci in a rich bacterial growth medium.
Anti-pneumococcus serum prepared from the non-virulent rough strain was added.

The living rough bacteria which he added to the broth could not survive in the presence of this serum. After several growth cycles in a test tube with the serum, pneumococci of the smooth type were detected. The transformation had worked without the need to inject the components into the mouse. The way was clear to purify the factor active in transformation, which they called "transforming principle". The highly purified transforming principle was DNA.

Avery's conclusions were published together with McCarty and MacLeod in a muted paper in 1944. For years, two influential scientists at Avery's own institution, Rockefeller Institute, questioned Avery's conclusions. P.A.T Levene's Tetranucleotide Hypothesis was incompatible with an informational role for DNA. And Alfred Mirsky, expert on chromatin, maintained that protein contaminants of Transforming Principle could be the true transforming agents – not DNA!

Avery's reticence raised doubts that he recognized the importance of his own findings. For many, however, the identification of the transforming principle as DNA was an exhilirating discovery.

WHAT IS DNA?

DNA is like a long string of beads in which each bead can be one of four kinds. Information is coded in the order in which the beads are arranged in the string. Each bead is known as a "base" and has a name: **adenine, guanine, cytosine** or **thymine**. (In RNA, whose function we will study later, thymine is replaced by **uracil**.) By 1900 all of these bases were known to chemists, and were classified into two groups: the **purines**, adenine and guanine; and the **pyrimidines**, cytosine, thymine and uracil. These are abbreviated A, G, C, T, and U.

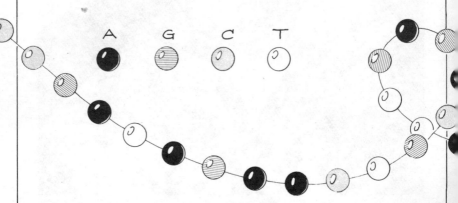

A   G   C   T

If DNA carried genetic information, the ratio of the bases would probably vary. If there were exactly the same number of adenines as guanines, cytosines and thymines, DNA might not carry information.

In the early twentieth century, chemists did not know that the genetic information resides in the **linear** arrangment of bases. Using rather crude chemical analyses, scientists were misled to believe that DNA contained exactly **equal** amounts of the four bases. This is called the Tetranucleotide Hypothesis which originated with a German chemist, Albrecht Kossel, but it is more closely associated with the name of a Russian born chemist who did most of his work at the Rockefeller Institute in New York, P.A.T. Levene.

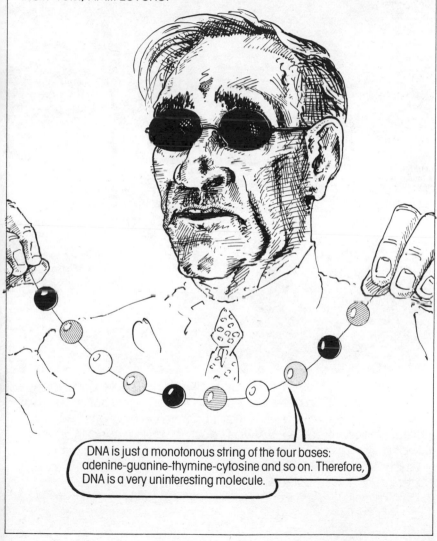

DNA is just a monotonous string of the four bases: adenine-guanine-thymine-cytosine and so on. Therefore, DNA is a very uninteresting molecule.

Levene has been much maligned for accepting the Tetranucleotide Hypothesis and even blamed for holding back genetic research for several decades as a consequence! Let's see how history has proved Levene to have been mistaken...

There are two different kinds of nucleic acids – DNA and RNA. Since, in the late nineteenth century, the primary source of what would be later called DNA was calf thymus, and what would later be identified as RNA was yeast, it was believed that DNA was only found in animals and RNA was only found in plants (such as yeast). This was an unfortunate misapprehension and it persisted in spite of the fact that RNA was frequently found in animal cells. Scientists dismissed this fact arguing that the animals had ingested plants!

At the chemical level, DNA and RNA are distinguished by: 1) that the base thymine is in DNA and is replaced by uracil in RNA; 2) that the sugar "backbones" are different. In both sugars there are five carbons arranged in a five sided ring structure. Such sugars are called pentoses, specifically riboses.

RNA is an abbreviation of $\underline{R}$ibo $\underline{N}$ucleic $\underline{A}$cid, and DNA is $\underline{D}$eoxyribo $\underline{N}$ucleic $\underline{A}$cid.

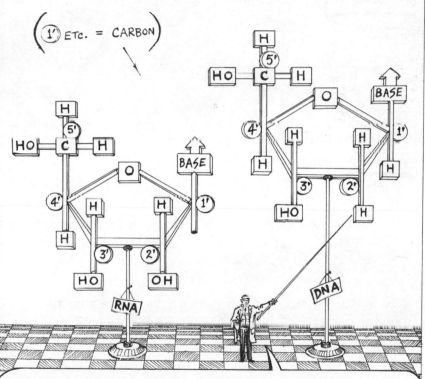

We give each sugar carbon a number; you will notice in the models that the number 2' carbon atom has a different side group in the DNA sugar as opposed to the RNA sugar. The DNA sugar lacks an oxygen atom at position number two. Since the RNA sugar is called "Ribose" the DNA sugar is called 2'**Deoxy**ribose, that is, a Ribose without oxygen at atom number 2'. **That is the origin of the abbreviated forms DNA and RNA.**

By 1935 Levene showed that the sugars are connected to each other in both DNA and RNA by a phosphodiester bond, that is, through a phosphorus surrounded by oxygens:—

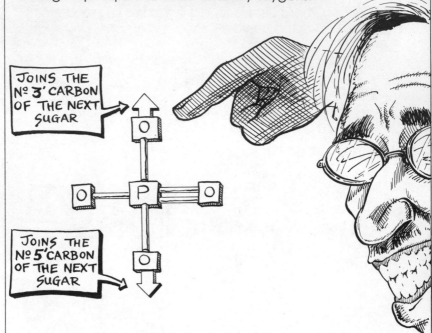

JOINS THE Nº 3' CARBON OF THE NEXT SUGAR

JOINS THE Nº 5' CARBON OF THE NEXT SUGAR

When attached together, a sugar and base are called a nucleotide. The bases (A, C, G, T or U) are attached to the number 1' carbon of the sugar.

Notice that the linkage of one sugar to the next is from the number 5' carbon, via a phosphodiester bond, to the number 3' carbon of the next sugar. This is why DNA is like a long set of beads on a string. The beads on the string are the sugars, each with an attached base. The string itself is the phosphodiester bond between sugars.

Because the phosphate linkage between sugars runs from the number five carbon to the number three (as shown on the next page), we can say that the sugar backbone can be orientated in space. Biochemists talk of "moving in the 5' to 3' direction" down a DNA chain.

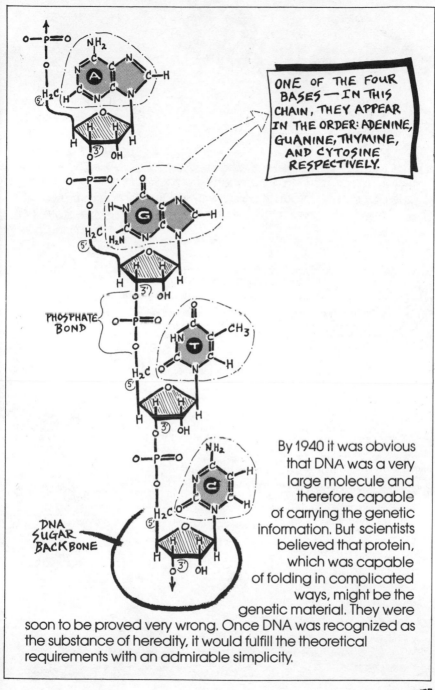

ONE OF THE FOUR BASES — IN THIS CHAIN, THEY APPEAR IN THE ORDER: ADENINE, GUANINE, THYMINE, AND CYTOSINE RESPECTIVELY.

PHOSPHATE BOND

DNA SUGAR BACKBONE

By 1940 it was obvious that DNA was a very large molecule and therefore capable of carrying the genetic information. But scientists believed that protein, which was capable of folding in complicated ways, might be the genetic material. They were soon to be proved very wrong. Once DNA was recognized as the substance of heredity, it would fulfill the theoretical requirements with an admirable simplicity.

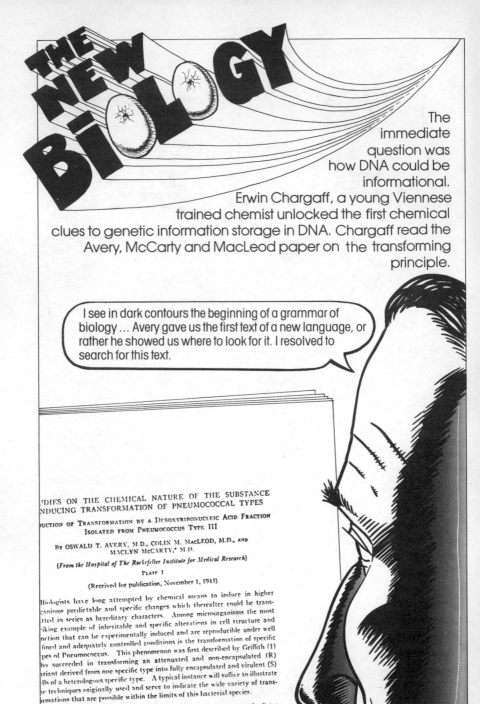

THE NEW BIOLOGY

The immediate question was how DNA could be informational. Erwin Chargaff, a young Viennese trained chemist unlocked the first chemical clues to genetic information storage in DNA. Chargaff read the Avery, McCarty and MacLeod paper on the transforming principle.

I see in dark contours the beginning of a grammar of biology … Avery gave us the first text of a new language, or rather he showed us where to look for it. I resolved to search for this text.

'DIES ON THE CHEMICAL NATURE OF THE SUBSTANCE
NDUCING TRANSFORMATION OF PNEUMOCOCCAL TYPES

DUCTION OF TRANSFORMATION BY A DESOXYRIBONUCLEIC ACID FRACTION
ISOLATED FROM PNEUMOCOCCUS TYPE III

BY OSWALD T. AVERY, M.D., COLIN M. MacLEOD, M.D., AND
MACLYN McCARTY,* M.D.

(From the Hospital of The Rockefeller Institute for Medical Research)

PLATE 1

(Received for publication, November 1, 1943)

Biologists have long attempted by chemical means to induce in higher
ganisms predictable and specific changes which thereafter could be trans-
itted in series as hereditary characters. Among microörganisms the most
riking example of inheritable and specific alterations in cell structure and
nction that can be experimentally induced and are reproducible under well
fined and adequately controlled conditions is the transformation of specific
pes of Pneumococcus. This phenomenon was first described by Griffith (1)
ho succeeded in transforming an attenuated and non-encapsulated (R)
ariant derived from one specific type into fully encapsulated and virulent (S)
lls of a heterologous specific type. A typical instance will suffice to illustrate
he techniques originally used and serve to indicate the wide variety of trans-
ormations that are possible within the limits of this bacterial species.

that mice injected subcutaneously with a small amount of a living

36

Chargaff's approach was to use the methods of quantitative analysis, bolstered by newly available techniques for separating the four bases. He purified DNA samples and then carefully quantified the amount of the four bases, A, G, C, and T.

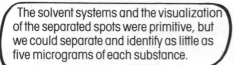

The solvent systems and the visualization of the separated spots were primitive, but we could separate and identify as little as five micrograms of each substance.

When Chargaff measured the base compositions of DNA from many sources he noted regularities summarized in "Chargaff's Rules":

(a) THE QUANTITY OF A + G = " " " C + T.

(b) A CONTENT ALWAYS = T CONTENT
    G " " = C "

This numerology was at once intriguing and enigmatic. It suggested that there was an underlying regularity to the composition of DNA. But Chargaff's rules on their own were insufficient to explain the regularities that Chargaff observed.

DNA is a macromolecule and most of its interesting features are lost when it is degraded. The successful experimental approach for determining DNA structure had to be capable of analyzing DNA intact, in its macromolecular form. One such technique was X-ray diffraction, which was developed in Cambridge by the Braggs, father and son, at the Cavendish laboratories.

SIR LAWRENCE BRAGG
CAVENDISH PROFESSOR
1938 – 1953

In X-ray diffraction, a fine beam of X-rays is passed through a crystal of the substance whose structure is under analysis. It interacts with the atoms in the crystal, and re-emerges as a complex pattern of beams that may be captured on X-ray film. By analyzing changes in the beam imparted by the specimen, the structure of the unknown molecule may be deduced.

The Cambridge lab sought a daring application of X-ray diffraction to the very complex biological macromolecules, the proteins. Max Perutz, an Austrian, was enlisted to lead the Cambridge protein structure team.

James D. Watson

Perutz's team did not consist solely of investigators interested in proteins. The Cambridge lab was joined in 1951 by an American, James D. Watson, who had other questions on his mind. It was Watson's interest in genes which led him to the Cavendish. As an undergraduate at the University of Chicago, Watson divided his time between bird watching and musing about biology. He had the good fortune of studying with Salvador Luria, a founder of the phage group, from whom he learned the principles of phage genetics.

Watson appreciated that DNA was the molecular key to genetics, but his weak background in chemistry limited his ability to understand genetic phenomena in terms of the chemical properties of the DNA molecule. On completing studies with Luria, Watson journeyed to Copenhagen, where he persued phage studies. Watson enjoyed journeying from his Copenhagen laboratory base, and in 1951 he travelled to a conference in Naples, where he chanced to meet Maurice Wilkins, a London crystallographer interested in DNA.

At the Naples meeting, Wilkins briefly showed an X-ray photo of DNA. Unlike a traditional photograph obtained with a camera lens and daylight, simply looking at the X-ray photo did not disclose the structure of the DNA molecule. Even the sharp spots easily obtained with simple mineral specimens were missing.

However, the presence of a regular, if fuzzy, geometric pattern in the X-ray photo confirms that the DNA sample is, at least partially, crystalline.

I would conclude that genes must have some regularity to their structure which would allow them to pack together in a nearly crystalline arrangement. The regularity could simplify the deduction of the structure of a gene.

Inspired by the X-ray photo, Watson sought a lab where he too could delve into the chemical structure of DNA. Watson arranged a shift from Copenhagen to Cambridge where he joined Max Perutz's group. Protein crystallography did not come easily to Watson, and he soon found himself preoccupied by conversations with Francis Crick.

42

Francis Crick was a 35 year old physicist who was developing the mathematics of X-ray diffraction for application to macromolecules. At heart, Crick was a theorist, trained in physics but drawn to biology by a fascination for understanding the activities of living things through...

The spatial distribution of their constituent atoms . . . the chemical physics of biology.

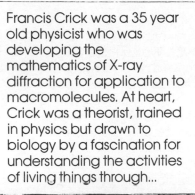

Crick studied physics until the outbreak of World War II, and then served in the Admiralty designing ingeneous magnetic mines. In 1949, he moved to the Cavendish group. Crick soon taught himself X-ray diffraction theory, and the current state of the protein structure problem.

From their first encounter in 1951, Watson and Crick thrived on each other's discussions. They agreed that the solution of DNA structure was the paramount problem of genetics. But Crick the theorist, and Watson the untutored newcomer, could contribute little new information of their own.

Outside Cambridge; one other scientist had novel insight into protein structure. Linus Pauling of Cal Tech, chemistry genius, proposed that protein chains fold in helical form. Crick knew this model well, and from studying Pauling's proposals he learned the theory of diffraction of helical macromolecules.

The best X-ray data, in part presented at the Naples meeting by Maurice Wilkins, resided in London. Wilkins was also a physicist who turned to biophysics, and in 1950, together with his graduate student Raymond Gosling, Wilkins obtained good X-ray patterns from DNA fibers.

Generally, the most definitive X-ray patterns are obtained when the specimen is crystalline. DNA, a long thread-like molecule, could be pulled into fibers, in which individual DNA molecules oriented themselves side by side, stretched out parallel to one another. Although not truly crystalline, the DNA fibers had sufficient order that informative X-ray patterns could be obtained. In forming the first fibers, most of the water was removed from the DNA, and the resulting structure was called "the A form" of DNA.

Towards the end of 1950, Wilkins was joined by Rosalind Franklin, an English, Cambridge-trained scientist, who had learned the theory and practice of X-ray diffraction in Paris. Franklin's introduction to X-ray work was with para-crystalline substances such as graphite. Thus, Franklin was well prepared to attack the problem of DNA structure which she found waiting when she arrived in London.

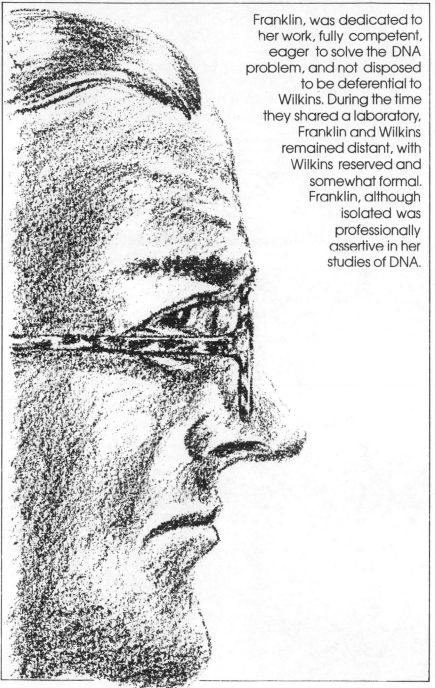

Franklin, was dedicated to her work, fully competent, eager to solve the DNA problem, and not disposed to be deferential to Wilkins. During the time they shared a laboratory, Franklin and Wilkins remained distant, with Wilkins reserved and somewhat formal. Franklin, although isolated was professionally assertive in her studies of DNA.

By the autumn of 1951, Franklin had her first success. She devised an improved method for adding water back to the A form of DNA fibers. When hydrated, the DNA underwent a dramatic structural change, observable by the diffraction technique.
In November 1951, Franklin gave the first public presentation of her results to a small gathering at Kings College. In the audience of this seminar was Watson.

The results suggest a helical structure (which must be very closely packed) containing probably 2, 3 or 4 co-axial nucleic acid chains per helical unit, and having the phosphate groups near the outside.

Naturally I was delighted when Maurice said I would be welcome at Rosy's talk. For the first time I had a real incentive to learn some crystallography. I did not want Rosy to speak over my head.

After the meeting, Watson and Wilkins had a Chinese dinner together. Watson left with the impression that Franklin had only refined Wilkin's existing data, and might actually slow the investigation because of her distant relationship with Wilkins. Watson returned to Cambridge and related his recollection of Franklin's talk to Crick.

Crick was tantalized by the possibility that the data already available might limit the structures for DNA to a small number of possibilities. The structure might be deduced by proposing a hypothetical structure, and fitting the experimental data against the predictions of the model.

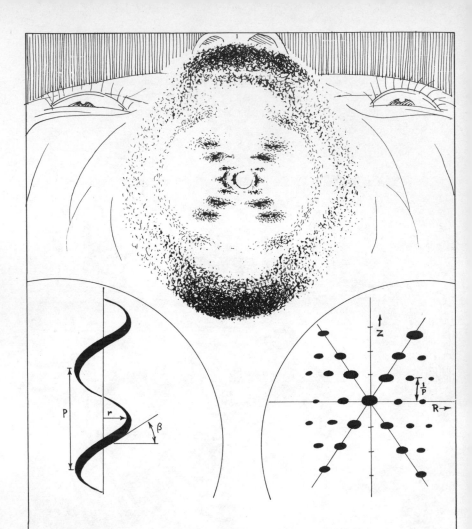

The X-ray diffraction pattern was in the form of a "maltese cross" characteristic of a helical molecule. Franklin and Wilkins had already recognized the probable helicity of DNA. Fortunately for Crick, the alpha helix protein structure proposed by Pauling had spurred an intense review and re-derivation of the theory of helical molecule diffraction, which Crick made together with Cochran. There were many possible helices: two stranded, three stranded, four stranded – and each could wind with a range of pitches and diameters. Many of the fundamental dimensions could be deduced from the X-ray photos.

In 1952 Erwin Chargaff traveled to Cambridge, and at the insistence of Perutz's co-worker, John Kendrew, he spoke to two people, Watson and Crick, at the Cavendish Laboratory who were "trying to do something with nucleic acids." Chargaff wrote of that meeting . . .

it there. I have to point out that mytholog-ical or historical couples - Castor and Pollux, Harmodios and Aristogeiton, Romeo and Juliet - must have appeared quite differently before the deed than after. In any event I seem to have missed the shiver of recognition of a historical moment: a change in the rhythm of the heart-beats of biology. So far as I could make out they wanted, unencumbered by any knowledge of the chemistry involved, to fit DNA into a helix.

"The high point of Chargaff's scorn", wrote James D. Watson of this meeting, "came when he led Francis into admitting that he did not remember the chemical difference among the four bases." Despite the scorn, Chargaff related in detail his findings about DNA base ratios. He described the numerology of Chargaff's Rules.

I am undeterred by Chargaff's scorn. But each visitor brings word of new facts about DNA. We must solve its structure!

Great! Let's build another model...

DID CHARGAFF SAY A ATTRACTS G OR G ATTRACTS T?

A difficult test of any model would be an explanation of the novel ability of genes to duplicate themselves. Watson was aware of a hypothesis that gene duplication relied upon formation of "complementary surfaces", from which a new gene could be constructed.

The mechanism would be similar to preparing a mould of an object, from which a replica of the original object could be cast.

An alternative scheme for duplication was direct copying, with no complementary intermediate.

Meanwhile, Pauling had turned his attention from proteins and proposed an unworkable structure for DNA. Watson recognized that Pauling's model failed to account for the acidic nature of DNA. Stabilizing forces, critical to Pauling's model, probably didn't exist. Watson was convinced that Pauling would soon be aware of his error, and then would intensify his effort to derive the correct structure.

During this time, Franklin pressed forward the X-ray studies of the B form of DNA. Watson was privy to her progress by virtue of periodic meetings with Maurice Wilkins in London. When the Cochran-Crick diffraction theory was used to test the B form patterns, it was evident that hydrated DNA was also a helix.

To bring together disparate evidence for the structure, Watson wanted to build precise representations of DNA helices. Machinists at the Cavendish were asked to make metal replicas of the purine and pyrimidine bases. In some models, the phosphate backbone was on the interior, as with Pauling's model. In others, it was on the exterior. Watson tried groupings of bases which could form hydrogen bonds and stabilize the helix. At one point, Watson considered a model in which A paired with A, T with T and so on, such that "like paired with like".

However, the like-like model was soon rejected because Watson had employed the wrong chemical forms of T and G.

Watson continued to shuffle the cut-outs of the bases in different combinations.

Suddenly I became aware that an adenine-thymine pair was identical in shape to a guanine-cytosine pair held together by at least two hydrogen bonds. All the hydrogen bonds seemed to form naturally, no fudging was required to make the two types of base pairs identical in shape.

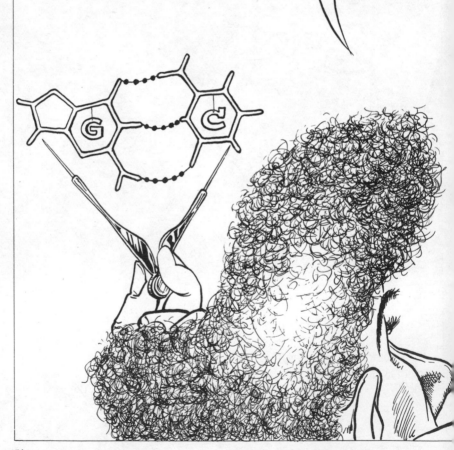

Most important was that the two types of base pairs had the same overall size and shape. Thus, in fitting these pairs into the helix, any order of AT's and GC's could be accommodated, and the same regular exterior phosphate backbone could be maintained.

Together, Watson and Crick assembled a three dimensional structure of a double helix which contained the newly conceived base pairs and employed the dimensions derived from X-ray measurements.

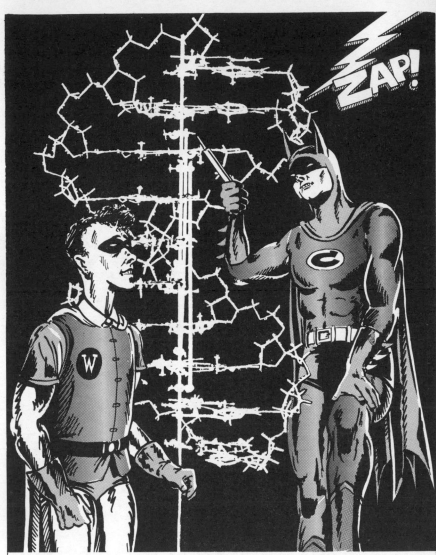

Standing as high as a man, the model had brass bases and wire sugars, and was held together by screws. Strikingly, the model at once suggested a mechanism for replicating genes. The base sequence of one chain automatically determined the sequence of the other. Also, the AT and GC base pairs immediately explained Chargaff's rules. With a C facing every G, and a T opposite every A, the A=T and G=C equivalences were assured.

The model was submitted to Nature in a 900 word manuscript, together with two other separate reports, one from Wilkins and one from Franklin. Watson's sister typed the final draft of the historic manuscript, which was sub-mitted in April, 1953, when Watson was 25 years old.

...Watson & Crick's model of DNA suggested the mechanism of replication.

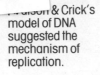

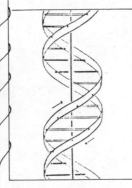

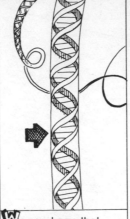

We now know that DNA synthesis begins at a **Replication Origin**.

One enzyme **unwinds** the double helix at the origin forming a **Replication Fork**.

At the fork the two separated strands serve as templates for new DNA synthesis.

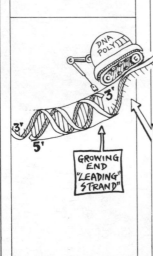

DNA POLY

3'
5'

GROWING END "LEADING" STRAND"

REPLICATION FORK

RNA PRIMER

Here are more enzymes called **DNA Polymerase** — they travel along the strands catalyzing the addition of DNA bases...

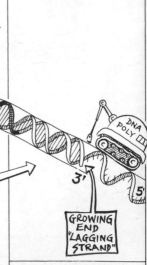

DNA POLY

3'
5'

GROWING END "LAGGING STRAND"

...to create two double strands.

Since Adenine always pairs with Thymine, & Cytosine always pairs with Guanine (the four bases), each **new** chain will be **complementary** to the parent chain that it uses as a template.

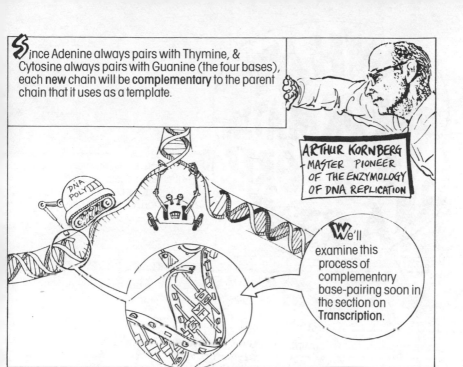

ARTHUR KORNBERG — MASTER PIONEER OF THE ENZYMOLOGY OF DNA REPLICATION

DNA POLY III

We'll examine this process of complementary base-pairing soon in the section on **Transcription.**

The interaction of these and many other features near the replication fork results in two new double helices. Each one has **one** chain from the original DNA molecule & one chain that has been newly formed.

HERE'S THE FIRST REPLICATION FORK WITH THE TWO NEW HELICES BELOW IT...

AND HERE'S A SECOND FORK WHERE ONE OF THE NEW HELICES IS REPLICATING

Here's a representation of what DNA replication looks like when hugely magnified:

WHAT INFORMATION IS STORED IN A GENE?

Replication is an extremely complicated process, but this guarantees the accuracy of genetic transmission, & consequently, life itself!

As we have seen **George Beadle** & **Edward Tatum** proposed:—

(SEE PAGE 25)

Each gene corresponds to one enzyme, or more precisely to one protein.

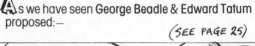

The critical clue to the complexity of this genetic information came from **Fred Sanger**, an English biochemist, who determined the complete amino acid sequence of the hormone, insulin.

INSULIN IS A PROTEIN. PROTEINS ARE LONG CHAINS OF AMINO ACIDS. THERE ARE TWENTY AMINO ACID TYPES.

Sanger proved that proteins have specific structures.

Sanger sequenced insulin by specifically degrading it into short fragments which were separated by a procedure known as **"fingerprinting"**.

TRYPSIN ENZYME (CLEAVES PROTEIN)

INSULIN SOLUTION

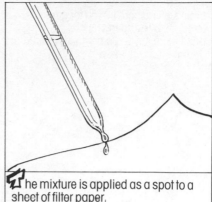

The mixture is applied as a spot to a sheet of filter paper.

Solvent is passed in one direction & electric current in the perpendicular direction.

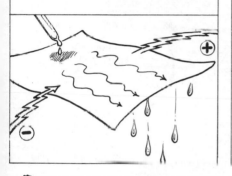

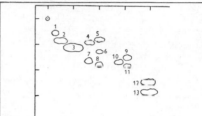

Depending on their solubility & electric charge, different fragments are moved to different positions on the paper, creating a distinct pattern.

When Sanger inadvertently touched the paper sheets before visualizing the protein fragments, spots appeared that were caused by protein from his fingertips...

SORRY 'BOUT THAT

Hence he called the patterns "fingerprints".

INSULIN

Also, like fingerprints, the patterns were characteristic for each protein: simple & reproducible, Sanger concluded that insulin had a specific structure. He next reassembled the short sequences into longer ones, & deduced the complete structure of insulin.

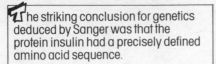The striking conclusion for genetics deduced by Sanger was that the protein insulin had a precisely defined amino acid sequence.

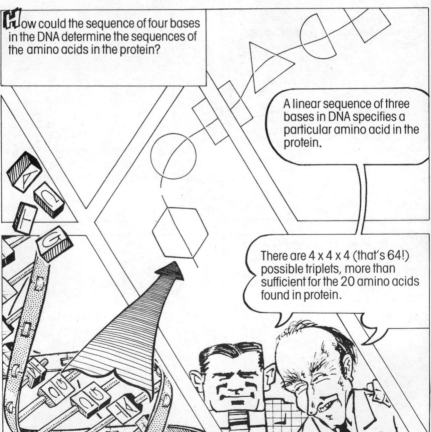

Thus the mechanism which directed the order of assembly of the individual amino acids of insulin was far from random &, itself, must have employed precisely defined instructions.

How could the sequence of four bases in the DNA determine the sequences of the amino acids in the protein?

A linear sequence of three bases in DNA specifies a particular amino acid in the protein.

There are 4 x 4 x 4 (that's 64!) possible triplets, more than sufficient for the 20 amino acids found in protein.

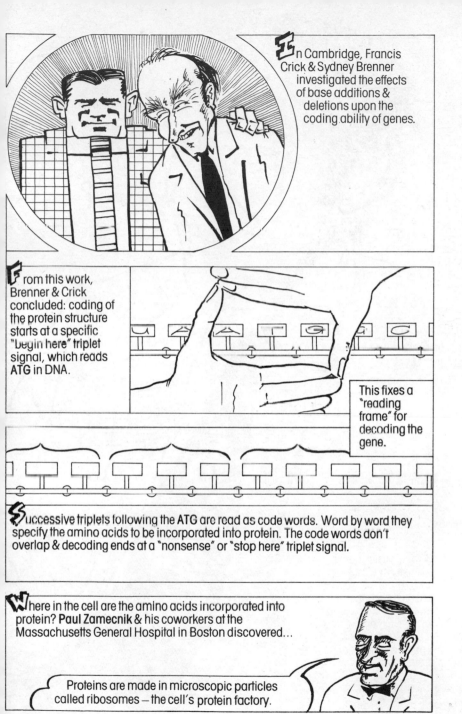

In Cambridge, Francis Crick & Sydney Brenner investigated the effects of base additions & deletions upon the coding ability of genes.

From this work, Brenner & Crick concluded: coding of the protein structure starts at a specific "begin here" triplet signal, which reads ATG in DNA.

This fixes a "reading frame" for decoding the gene.

Successive triplets following the ATG are read as code words. Word by word they specify the amino acids to be incorporated into protein. The code words don't overlap & decoding ends at a "nonsense" or "stop here" triplet signal.

Where in the cell are the amino acids incorporated into protein? **Paul Zamecnik** & his coworkers at the Massachusetts General Hospital in Boston discovered...

Proteins are made in microscopic particles called ribosomes — the cell's protein factory.

But what is happening inside the ribosome

Francis Crick had a brainstorm!! At the time he and a group of twenty other scientists had formed an elite **RNA Tie Club**. It was to this exclusive membership — and to them only — that Crick sent a mimeographed copy of his brainstorm:

THE ADAPTOR HYPOTHESIS

Crick's Adaptor Hypothesis dealt with the problem of how the code held in the DNA double helix gets **translated** into protein.

I KNOW (OR CAN ASSUME) THE FOLLOWING

The code in DNA is **linear**. Its sequence of bases is like a string of beads.

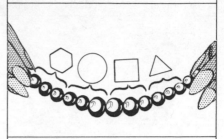

Every group of three bases in DNA codes for **one** amino acid.

A protein is made up from a string of amino acids (of which there are twenty different kinds).

The amino acids in a protein are in the same sequence as their codes in DNA.

What are the steps between DNA triplets and amino acids?

PART WHICH RECOGNISES A TRIPLET OF BASES

PART WHICH CARRIES ONE OF THE TWENTY AMINO ACIDS

There must be an **Adaptor** molecule which has one side to recognise one amino acid & the other to recognise its triplet code.

It wasn't long before scientists discovered the adaptors Crick had predicted. They were made out of RNA & **translated** the genetic code (which was stored in a molecule called **messenger RNA**) inside the **ribosome**. The ribosome was the assembly line for the production of proteins.

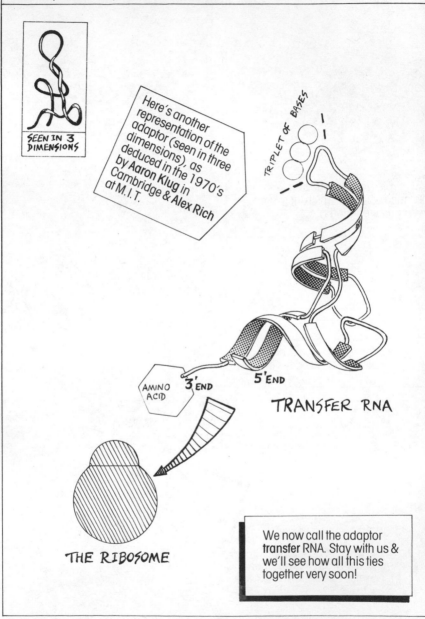

SEEN IN 3 DIMENSIONS

Here's another representation of the adaptor (seen in three dimensions), as deduced in the 1970's by **Aaron Klug** in Cambridge & **Alex Rich** at M.I.T.

TRIPLET OF BASES

AMINO ACID   3' END

5' END

TRANSFER RNA

THE RIBOSOME

We now call the adaptor **transfer** RNA. Stay with us & we'll see how all this ties together very soon!

Remember Zamecnik? He found…

Proteins are made in the ribosomes & there is **no** DNA in the ribosomes.

How did the coded information get from the cell's DNA to the ribosome?

It was only after **Arthur Pardee, Francois Jacob & Jacques Monod** working together at the Pasteur Institute in Paris performed their famous "**PaJaMo**" experiments (named after themselves), that this piece of the puzzle fell into place.

β-G.

(SEE PAGE 90)

The gene for making an enzyme, Beta Galactosidase (which digested the sugar, lactose) were transferred from the male bacteria to the females which were not capable of making the enzyme.

(As we shall see later, bacteria have "sex".)

(SEE PAGE 98)

69

The gene for β-Galactosidase no sooner entered the female than it (she) began producing the enzyme β-Galactosidase.

This surprised most scientists because it was widely assumed that before a gene could be expressed (before a cell could produce the protein for which the gene codes) **stable** cellular structures would have to form & accumulate.

This meant that there would be a delay before the gene products appeared in a bacterium.

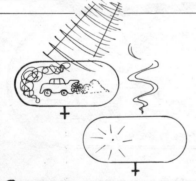

Other experiments were performed in which a newly inserted gene (DNA) was destroyed by a radioactive technique (which we needn't detail). Once the gene was destroyed the bacterium stopped producing the particular product encoded in that gene. This too was surprising.

After much discussion on Good Friday in 1960 in Cambridge, England, Crick, Brenner & Jacob concluded that these experiments showed that the template for protein synthesis (what the tRNA attached to) was **not stable**.

RHUBARB RHUBARB

They concluded that as soon as the male DNA entered the female bacterium **unstable, short-lived** copies of the DNA were formed, & these copies were the templates for protein synthesis

The centre for protein assembly — the ribosome — was known to be stable...

HO OH
HO H H

HO OH
O
HO OH

DNA SUGAR

RNA SUGAR

and tRNA was also known to be stable...

— so a **Messenger RNA** was postulated.

m RNA

RNA because:—
1. RNA & **not** DNA was found in the ribosomal protein factories;
2. An RNA copy of the DNA, generated by base-pairing, would store the same information as the DNA itself.

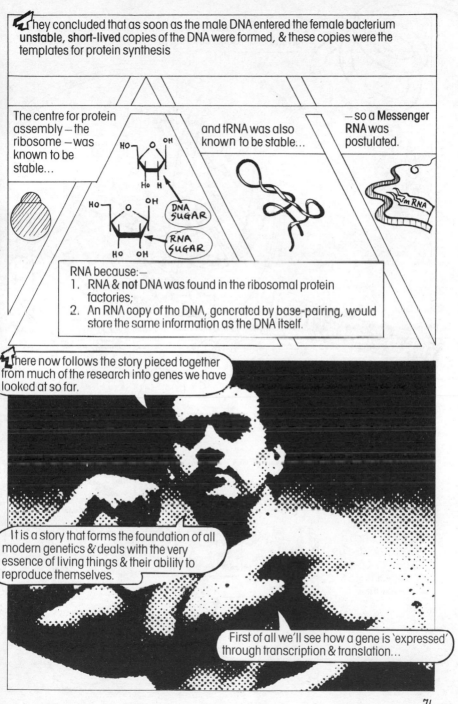

There now follows the story pieced together from much of the research into genes we have looked at so far.

It is a story that forms the foundation of all modern genetics & deals with the very essence of living things & their ability to reproduce themselves.

First of all we'll see how a gene is 'expressed' through transcription & translation...

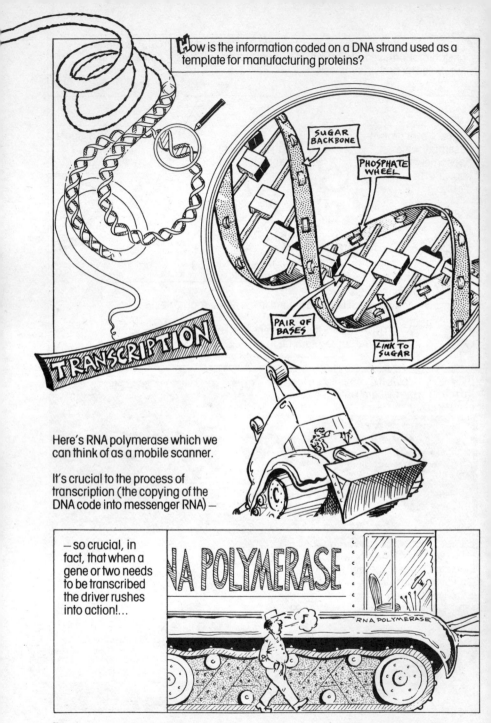

How is the information coded on a DNA strand used as a template for manufacturing proteins?

SUGAR BACKBONE

PHOSPHATE WHEEL

PAIR OF BASES

LINK TO SUGAR

TRANSCRIPTION

Here's RNA polymerase which we can think of as a mobile scanner.

It's crucial to the process of transcription (the copying of the DNA code into messenger RNA) —

— so crucial, in fact, that when a gene or two needs to be transcribed the driver rushes into action!…

NA POLYMERASE

RNA POLYMERASE

The mobile scanner senses the **Promoter** on the DNA and rolls into position on the **Initiation Site**, causing the strands to unwind!!...

...as it rolls forward over the strand to be transcribed.

...ANOTHER DAY, ANOTHER DOUBLE HELIX...

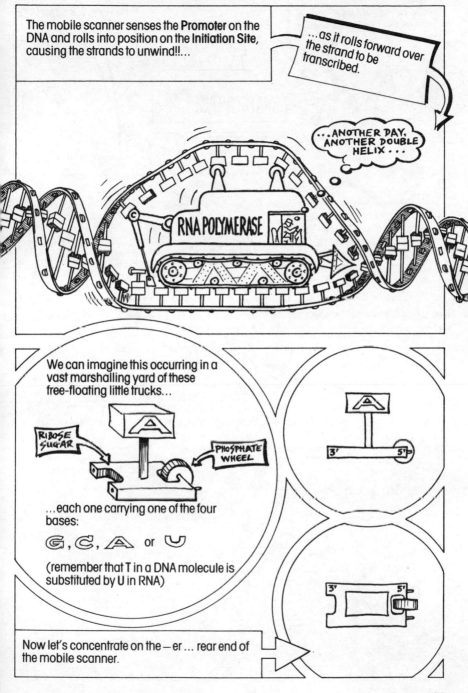

RNA POLYMERASE

We can imagine this occurring in a vast marshalling yard of these free-floating little trucks...

RIBOSE SUGAR

PHOSPHATE WHEEL

...each one carrying one of the four bases:

G, C, A or U

(remember that T in a DNA molecule is substituted by U in RNA)

Now let's concentrate on the — er ... rear end of the mobile scanner.

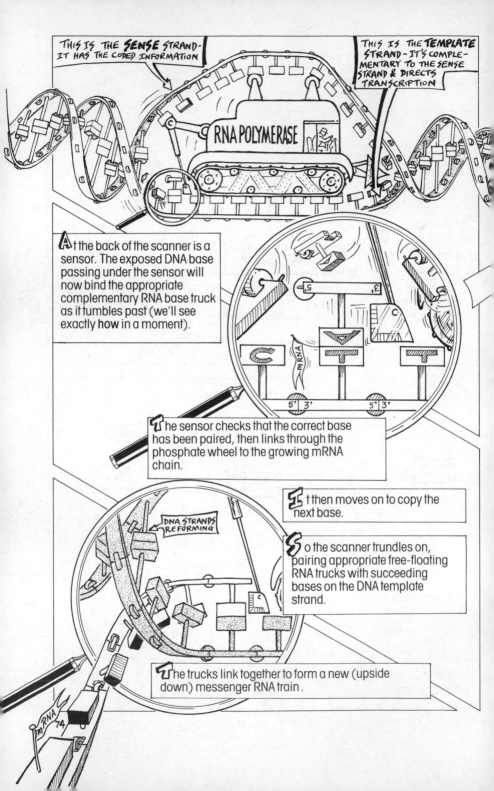

THIS IS THE **SENSE** STRAND - IT HAS THE CODED INFORMATION

THIS IS THE **TEMPLATE** STRAND - IT'S COMPLEMENTARY TO THE SENSE STRAND & DIRECTS TRANSCRIPTION

RNA POLYMERASE

At the back of the scanner is a sensor. The exposed DNA base passing under the sensor will now bind the appropriate complementary RNA base truck as it tumbles past (we'll see exactly **how** in a moment).

The sensor checks that the correct base has been paired, then links through the phosphate wheel to the growing mRNA chain.

It then moves on to copy the next base.

So the scanner trundles on, pairing appropriate free-floating RNA trucks with succeeding bases on the DNA template strand.

DNA STRANDS REFORMING

The trucks link together to form a new (upside down) messenger RNA train.

74

Here are the little devils as they pair in DNA, with (clearly visible) those hydrogen bonds holding the bases together.

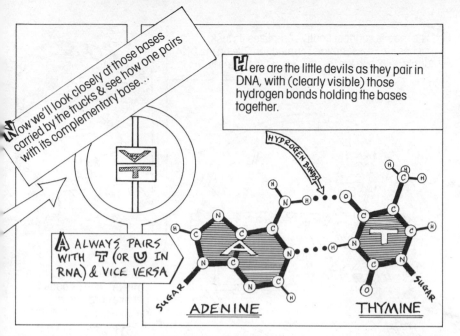

HYDROGEN BONDS

A ALWAYS PAIRS WITH T (OR U IN RNA) & VICE VERSA

ADENINE   THYMINE

During transcription the incoming RNA trucks pair precisely with the DNA template strand. Because this copy is complementary to the template, it has the same sequence of bases as the DNA sense strand and therefore contains the coded genetic information.

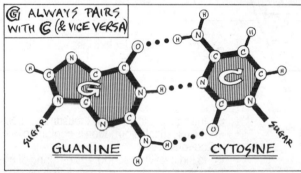

G ALWAYS PAIRS WITH C (& VICE VERSA)

GUANINE   CYTOSINE

Here's the structure of DNA
r = ribose sugar
p = phosphate

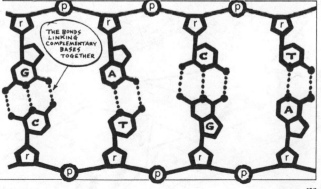

THE BONDS LINKING COMPLEMENTARY BASES TOGETHER

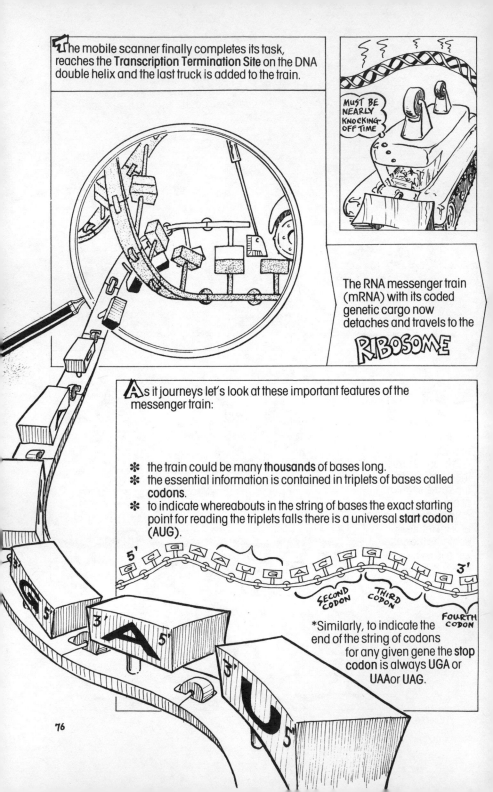

The mobile scanner finally completes its task, reaches the **Transcription Termination Site** on the DNA double helix and the last truck is added to the train.

MUST BE NEARLY KNOCKING OFF TIME

The RNA messenger train (mRNA) with its coded genetic cargo now detaches and travels to the

RIBOSOME

As it journeys let's look at these important features of the messenger train:

✱ the train could be many **thousands** of bases long.
✱ the essential information is contained in triplets of bases called **codons**.
✱ to indicate whereabouts in the string of bases the exact starting point for reading the triplets falls there is a universal **start codon** (AUG).

SECOND CODON   THIRD CODON   FOURTH CODON

*Similarly, to indicate the end of the string of codons for any given gene the **stop codon** is always UGA or UAA or UAG.

In the cells of prokaryotes — such as bacteria (that is cells **without** nuclei) — an mRNA can hold the codes for a number of genes:

We will see later that the genes & mRNA of the cells of eukaryotes — such as plants and animals (cells **with** nuclei) — are very different.

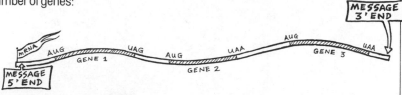

MESSAGE 3' END

mRNA

AUG   UAG   AUG   UAA   AUG   UAA

GENE 1   GENE 2   GENE 3

MESSAGE 5' END

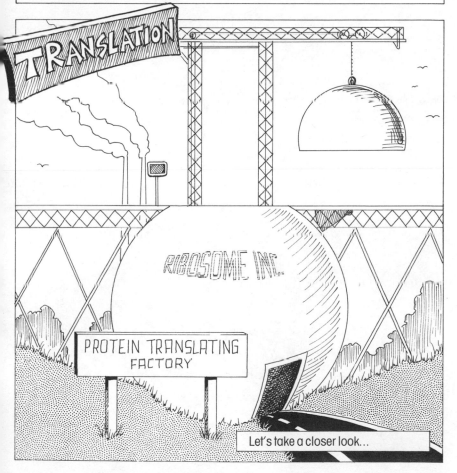

TRANSLATION

RIBOSOME INC.

PROTEIN TRANSLATING FACTORY

Let's take a closer look…

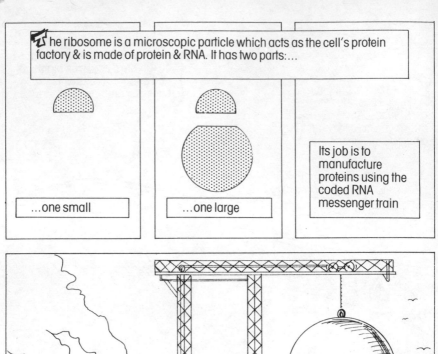

The ribosome is a microscopic particle which acts as the cell's protein factory & is made of protein & RNA. It has two parts:...

...one small

...one large

Its job is to manufacture proteins using the coded RNA messenger train

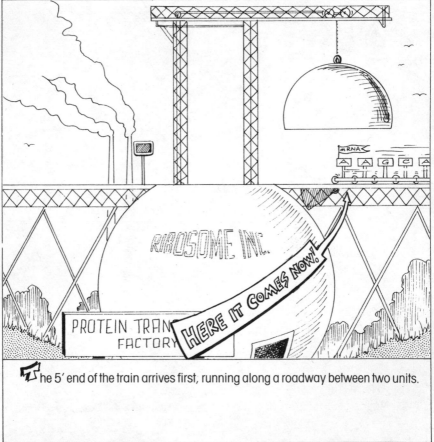

RIBOSOME INC.

mRNA

A A G C A

HERE IT COMES NOW!

PROTEIN TRAN... FACTORY

The 5' end of the train arrives first, running along a roadway between two units.

When the <u>start codon</u> passes under the small unit, it causes a gantry to lower that unit onto **AUG** & the **next codon**. At the same time an automatic ramp flips the train **upside down**!!

These two codons then move into position with the small unit above them, to occupy platforms **P & A** respectively in the large unit.

(In the bowels of the factory)

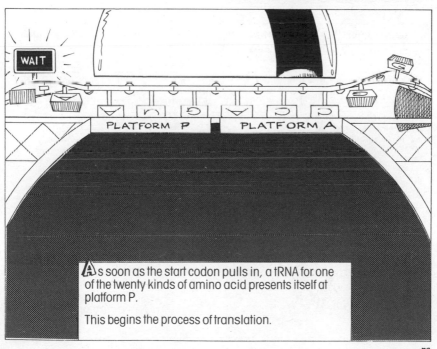

As soon as the start codon pulls in, a tRNA for one of the twenty kinds of amino acid presents itself at platform P.

This begins the process of translation.

JUST A MINUTE! tRNA?? WHAT'S THAT??!

Remember Crick's adaptor hypothesis? It pointed towards a small RNA molecule (some 70 to 90 bases long) in a clover leaf shape when seen in 2 dimensions. It is called transfer RNA (tRNA) as its job is to transfer amino acids from a free state to a growing protein chain. After all, that's why we're all here, isn't it?

It has two main parts: a head consisting of a triplet of bases (which is called the anti-codon as it complements the codon of the mRNA train); and a trailer at the rear which carries one of the twenty amino acids (the **components** which make up **proteins**)

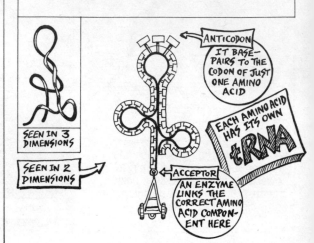

SEEN IN 3 DIMENSIONS

SEEN IN 2 DIMENSIONS

ANTICODON
IT BASE-PAIRS TO THE CODON OF JUST ONE AMINO ACID

EACH AMINO ACID HAS ITS OWN tRNA

ACCEPTOR
AN ENZYME LINKS THE CORRECT AMINO ACID COMPONENT HERE

We can imagine that beneath the protein factory is a vast underground warehouse containing many many tRNA's carrying amino acids.

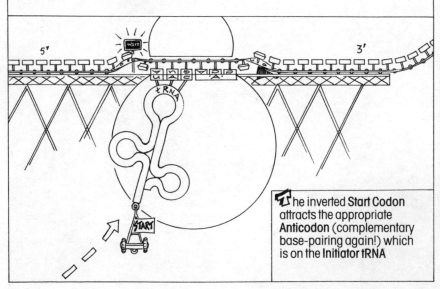

WAIT

5'   3'

tRNA

START

The inverted **Start Codon** attracts the appropriate **Anticodon** (complementary base-pairing again!) which is on the **Initiator tRNA**

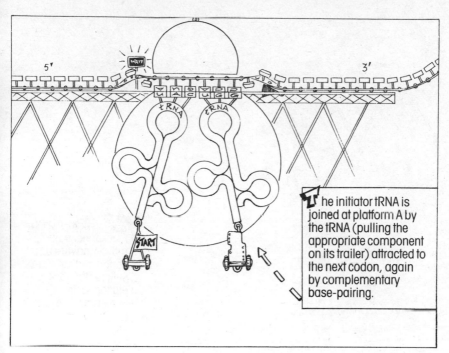

The initiator tRNA is joined at platform A by the tRNA (pulling the appropriate component on its trailer) attracted to the next codon, again by complementary base-pairing.

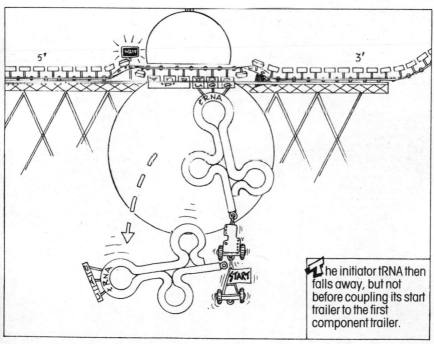

The initiator tRNA then falls away, but not before coupling its start trailer to the first component trailer.

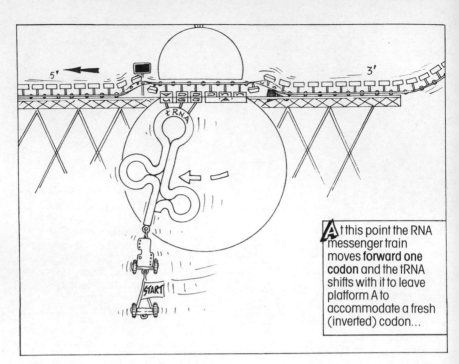

5′

3′

tRNA

START

At this point the RNA messenger train moves **forward one codon** and the tRNA shifts with it to leave platform A to accommodate a fresh (inverted) codon...

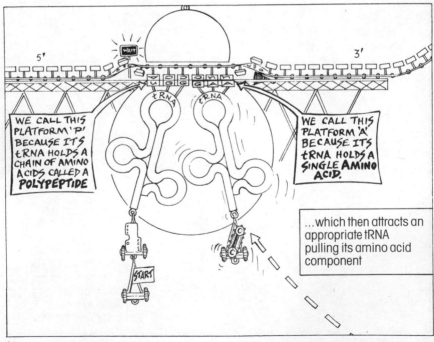

5′

3′

WAIT

tRNA   tRNA

WE CALL THIS PLATFORM 'P' BECAUSE ITS tRNA HOLDS A CHAIN OF AMINO ACIDS CALLED A **POLYPEPTIDE**

WE CALL THIS PLATFORM 'A' BECAUSE ITS tRNA HOLDS A SINGLE **AMINO ACID.**

START

...which then attracts an appropriate tRNA pulling its amino acid component

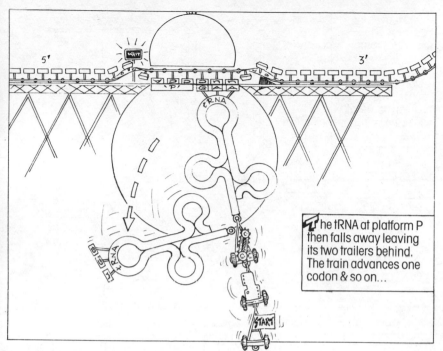

The tRNA at platform P then falls away leaving its two trailers behind. The train advances one codon & so on...

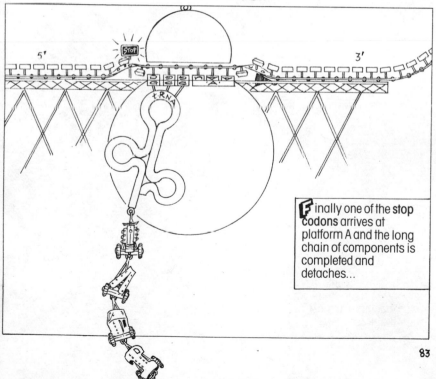

Finally one of the **stop codons** arrives at platform A and the long chain of components is completed and detaches...

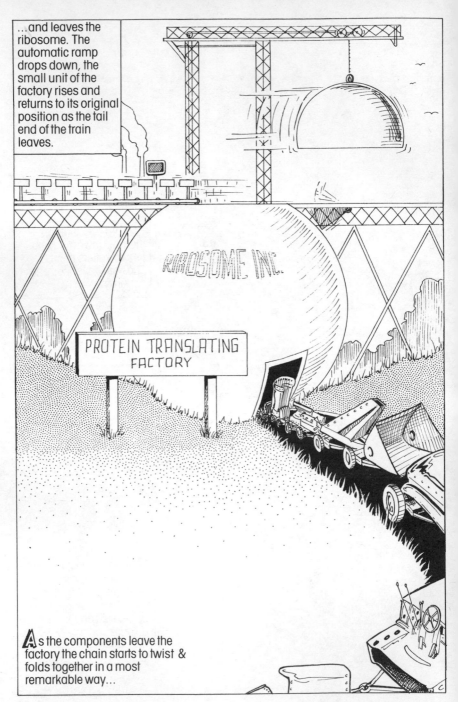

...and leaves the ribosome. The automatic ramp drops down, the small unit of the factory rises and returns to its original position as the tail end of the train leaves.

RIBOSOME INC.

PROTEIN TRANSLATING FACTORY

As the components leave the factory the chain starts to twist & folds together in a most remarkable way...

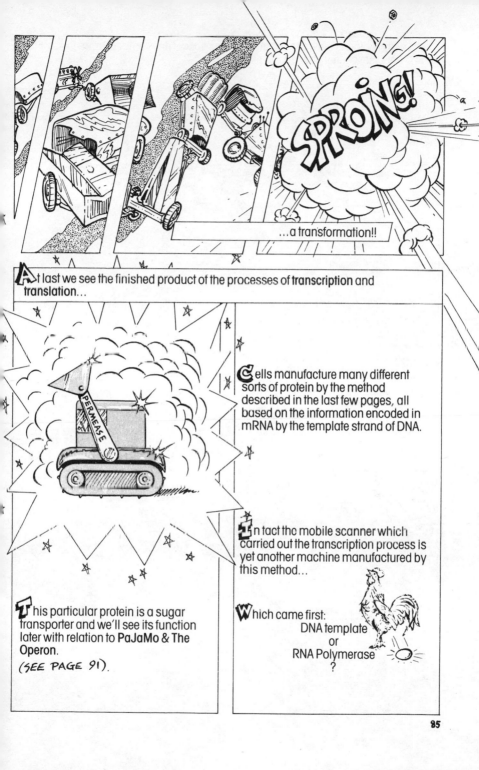

...a transformation!!

At last we see the finished product of the processes of **transcription** and **translation**...

Cells manufacture many different sorts of protein by the method described in the last few pages, all based on the information encoded in mRNA by the template strand of DNA.

In fact the mobile scanner which carried out the transcription process is yet another machine manufactured by this method...

This particular protein is a sugar transporter and we'll see its function later with relation to **PaJaMo & The Operon**.
(SEE PAGE 91).

Which came first:
        DNA template
            or
        RNA Polymerase
            ?

It's worth pointing out that there can be a number of ribosomes all translating the same RNA messenger train at different points.

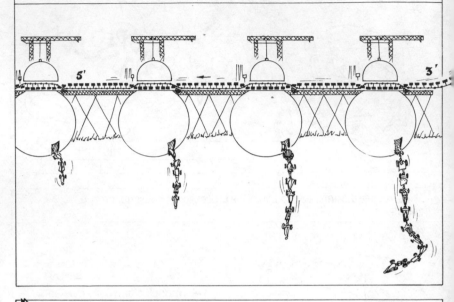

Also that the RNA messenger train is **short-lived** unlike the long-lived template DNA strand from which it was assembled. This means that when the train has passed through the last ribosome, its job having been done…

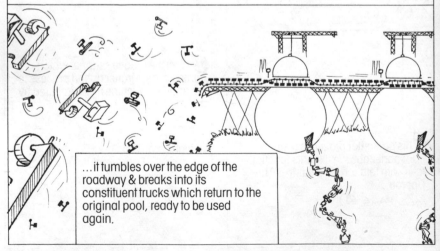

…it tumbles over the edge of the roadway & breaks into its constituent trucks which return to the original pool, ready to be used again.

The code itself was cracked in the early 1960's. **Severo Ochoa** at the New York University Medical School devised enzymatic methods for making RNA molecules in the test tube which had defined nucleotide sequences.

Marshall Nirenberg and his student Phil Leder working at the National Institutes of Health in Maryland used synthetic RNA made by Ochoa's methods to direct protein synthesis by cell extracts in the test tube.

Ochoa

Nirenberg.

They found that simple RNA trinucleotides, the minimal molecules for specifying a code word, were sufficient for binding tRNA to ribosomes. The RNA triplet would bind to the ribosome, and guide only one tRNA in place. In one typical experiment GUU was the added triplet, and only valine tRNA was bound to the ribosome. Thus, GUU is a code word for valine. Remember there are 64 code words but only twenty amino acids. Some amino acids can be specified by more than one triplet. Thus ACU, ACC, ACA and ACG all code for threonine. Only three triplets failed to direct tRNA binding: UAA, UAG and UGA. These are the "nonsense" or "stop here" signals postulated by Crick and Brenner.

GENETIC CODE CRACKED FULL STORY!

KEY

PHE – PHENYLALANINE
GLU – GLUTAMIC ACID
ASP – ASPARTIC ACID
ASPN – ASPARAGINE
ILEU – ISOLEUCINE
MET – METHIONINE
THR – THREONINE
ARG – ARGININE
GLUN – GLUTAMINE
HIS – HISTIDINE
TRP – TRYPTOPHAN
TYR – TYROSINE
CYS – CYSTEINE
LEU – LEUCINE
PRO – PROLINE
ALA – ALANINE
VAL – VALINE
GLY – GLYCINE
LYS – LYSINE
SER – SERINE

	U	C	A	G	
U	PHE PHE LEU LEU	SER SER SER SER	TYR TYR Ochre Amber	CYS CYS Opal TRP	U C A G
C	LEU LEU LEU LEU	PRO PRO PRO PRO	HIS HIS GLUN GLUN	ARG ARG ARG ARG	U C A G
A	ILEU ILEU ILEU MET	THR THR THR THR	ASPN ASPN LYS LYS	SER SER ARG ARG	U C A G
G	VAL VAL VAL VAL	ALA ALA ALA ALA	ASP ASP GLU GLU	GLY GLY GLY GLY	U C A G

Here it is. The code for each of the twenty amin
So simple isn't it? Read the table an
cant i?

88

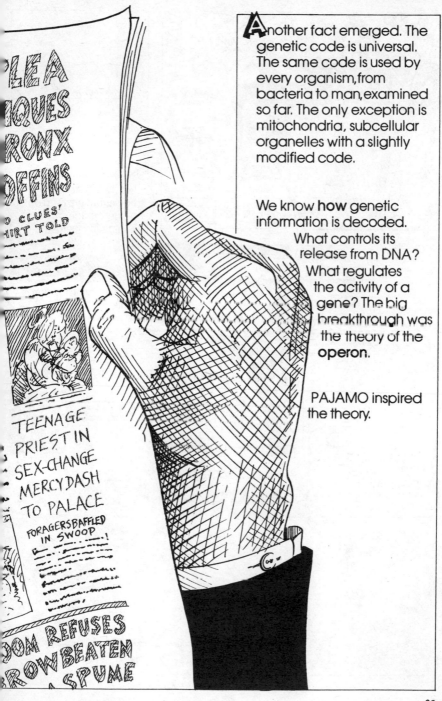

Another fact emerged. The genetic code is universal. The same code is used by every organism, from bacteria to man, examined so far. The only exception is mitochondria, subcellular organelles with a slightly modified code.

We know **how** genetic information is decoded. What controls its release from DNA? What regulates the activity of a gene? The big breakthrough was the theory of the **operon**.

PAJAMO inspired the theory.

But now we in the PaJaMo team have found **this** mutant bacterium which always produces the digestive enzyme even when there is **no** sugar present!

We have called this a **constitutive mutant**

We wonder about this...

Maybe the **regulation** of manufacture of the digestive enzyme is controlled by a gene that is **different** from the genes that are responsible for their structure...

Incidentally, we have since discovered how the digestive enzyme gets its supplies of sugar through the cell wall. This little fellow is called **Permease** because it helps sugar to permeate the cell wall!!

We're coming to that!

A bacterium with DNA inside…

…(made of some of the 4 million DNA base trucks)…

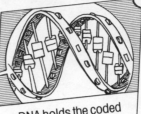

…DNA holds the coded genetic information…

…as we have seen, the coded information is transcribed by RNA Polymerase – the mobile scanner.

What determines whether digestive enyzme & permease are manufactured?

We conclude that the DNA contains a gene that codes for a **Repressor** molecule! This inhibits transcription of the digestive enzyme gene when there is no sugar present.

We'll call the sugar the **inducer** as it's only when it's present that (in the 'normal' bacterium) the repressor does not function & the manufacture of these little machines is induced.

CELL WALL

SUGAR

So we have two sets of genes – one set encodes proteins that regulate transcription…

…the other set determines the structure of proteins like digestive enzyme and permease.

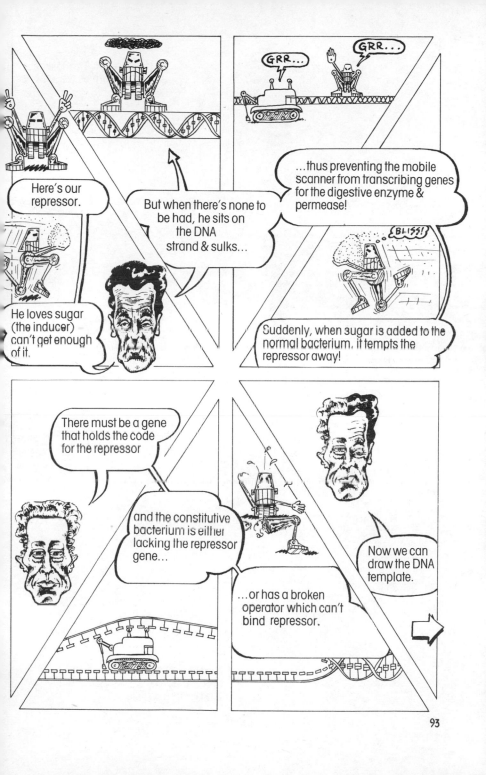

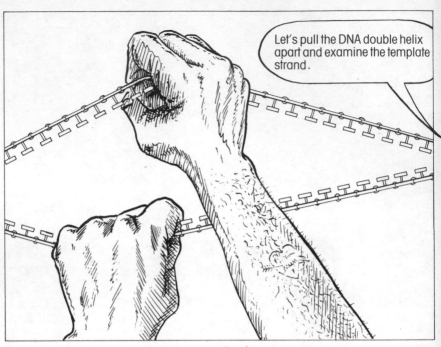

Let's pull the DNA double helix apart and examine the template strand.

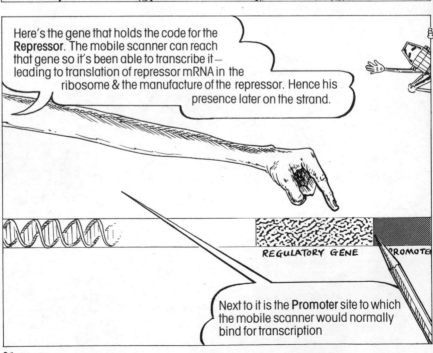

Here's the gene that holds the code for the **Repressor**. The mobile scanner can reach that gene so it's been able to transcribe it — leading to translation of repressor mRNA in the ribosome & the manufacture of the repressor. Hence his presence later on the strand.

REGULATORY GENE   PROMOTER

Next to it is the **Promoter** site to which the mobile scanner would normally bind for transcription

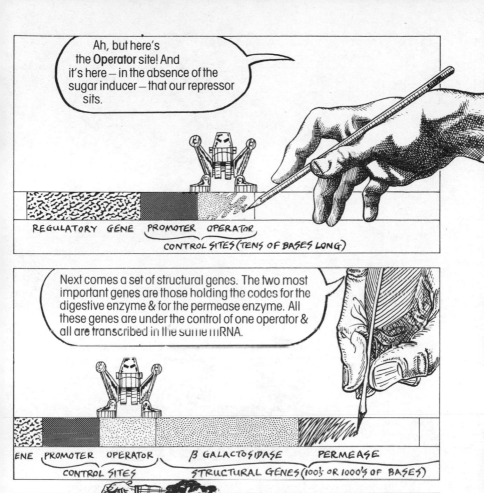

Ah, but here's the **Operator** site! And it's here — in the absence of the sugar inducer — that our repressor sits.

REGULATORY GENE   PROMOTER   OPERATOR

CONTROL SITES (TENS OF BASES LONG)

Next comes a set of structural genes. The two most important genes are those holding the codes for the digestive enzyme & for the permease enzyme. All these genes are under the control of one operator & all are transcribed in the same mRNA.

ENE   PROMOTER   OPERATOR   β GALACTOSIDASE   PERMEASE

CONTROL SITES   STRUCTURAL GENES (100's OR 1000's OF BASES)

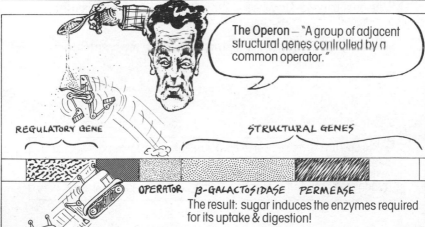

The Operon — "A group of adjacent structural genes controlled by a common operator."

REGULATORY GENE   STRUCTURAL GENES

OPERATOR   β-GALACTOSIDASE   PERMEASE

The result: sugar induces the enzymes required for its uptake & digestion!

he Operon model disclosed by PaJaMo ranks with
Crick's Adaptor Hypothesis...

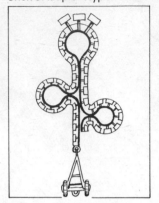
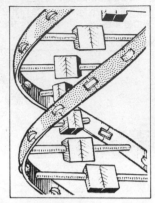

...and the Watson-Crick model as one
of the major intellectual achievements of modern biology!

AW, SHUCKS

IT WAS NUTHIN'

Operon control was only the first form of bacterial gene regulation to be discovered!

Since the 1960's many others, just as fascinating, have been found!

he operon provides diversity of gene expression for the **individual organism**, in response to hour by hour changes in the environment. But what created the diversity of the genes **themselves** that reside in different organisms? This is the diversity of variation and of speciation itself.

THE DIVERSITY OF GENE EXPRESSION

Mutation is one source of gene diversity.

But life on earth began some 3 to 4 billion years ago.

I DON'T SEEM TO BE GETTING VERY FAR

If the only cause of variation was random mutation, evolution would have been very slow!

Sexual reproduction could have provided great variability in primitive organisms. But can primitive organisms like bacteria have sex?

Bacterial sex seems absurd!

I'm puzzled by Avery's results

Well, **Joshua Lederberg** at 19 wondered about just that.

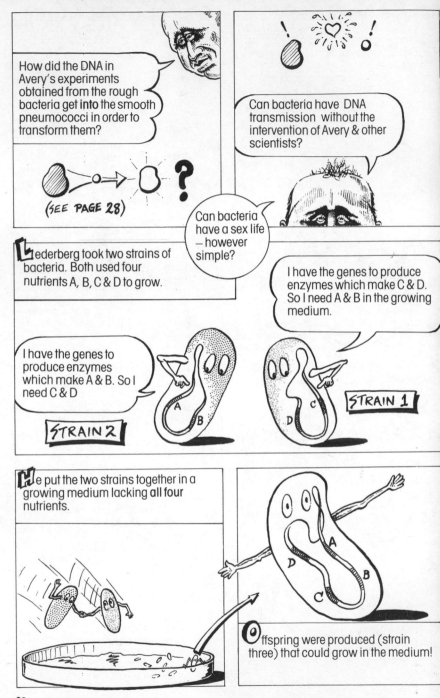

How did the DNA in Avery's experiments obtained from the rough bacteria get **into** the smooth pneumococci in order to transform them?

(SEE PAGE 28)

Can bacteria have DNA transmission without the intervention of Avery & other scientists?

Can bacteria have a sex life – however simple?

Lederberg took two strains of bacteria. Both used four nutrients A, B, C & D to grow.

I have the genes to produce enzymes which make C & D. So I need A & B in the growing medium.

I have the genes to produce enzymes which make A & B. So I need C & D

STRAIN 2

STRAIN 1

He put the two strains together in a growing medium lacking **all four** nutrients.

Offspring were produced (strain three) that could grow in the medium!

98

Because mutations are rare (one in a million) & because I am using bacterial strains 1 & 2, needing **two** nutrient substances each, the chances of getting bacteria which need none of these nutrients through mutation are practically zero (one million times one million)!

Strain three bacteria must have been produced through sexual conjugation (that is, transmission of DNA) between strains 1 & 2.

Bacteria have a sex life!

Having discovered that bacteria have a sex life, biologists soon found that they have **sexes** as well, and explained how genes were transferred from **1** to **2** to make **3**. Male bacteria (or what biologists have dubbed the males) have a piece of DNA called the **F-factor**, (fertility factor). Those bacteria that have the F-factor have special appendages called **sex-pili** which bind to receptor sites on the female bacterium (bacteria without the F-factor). A DNA molecule is transferred from the male to female through the sex-pili!

DNA from the male enters the female in a linear fashion. (Jacob called this the spaghetti hypothesis.) If A, B, C, D, E, F, G, ... Z are the genes in the male bacterium, first A, then B etc. will enter the bacteria in order and at a fixed rate of entry. They might enter starting anywhere in the sequence (say J), then move on to Z and continue through A to I. In this process, the F factor (DNA with genes for "maleness") may leave the male and enter the female, making the female a male and the former male a female! Maleness, as some biologists liked to say, is contagious!

Other pieces of DNA can enter a bacterium. One, a small DNA circle which remains separate from the bacterial chromosome, is called a plasmid. It's very useful in the cloning of genes, as we'll soon see.

PHOTO: COURTESY DR. CHARLES BRINTON JR.

Another way DNA can enter the bacterium is from

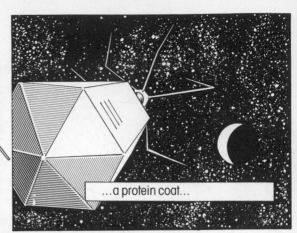

...a protein coat...

...enclosing pure DNA...

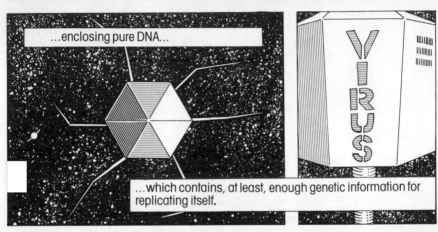

VIRUS

...which contains, at least, enough genetic information for replicating itself.

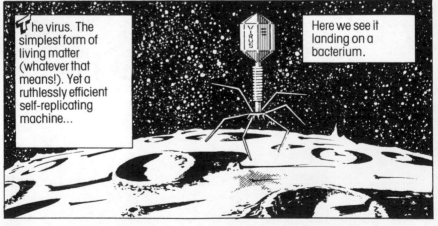

The virus. The simplest form of living matter (whatever that means!). Yet a ruthlessly efficient self-replicating machine...

VIRUS

Here we see it landing on a bacterium.

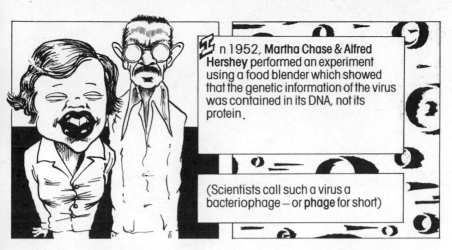

In 1952, **Martha Chase & Alfred Hershey** performed an experiment using a food blender which showed that the genetic information of the virus was contained in its DNA, not its protein.

(Scientists call such a virus a bacteriophage — or **phage** for short)

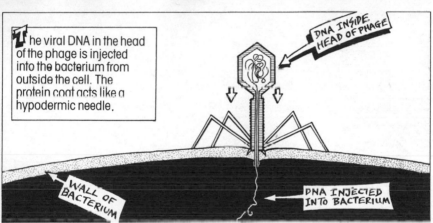

The viral DNA in the head of the phage is injected into the bacterium from outside the cell. The protein coat acts like a hypodermic needle.

DNA INSIDE HEAD OF PHAGE

WALL OF BACTERIUM

DNA INJECTED INTO BACTERIUM

Once the viral DNA is inside it can take control of the bacterial cell and by expressing its genes using the bacterium's mobile scanners, tRNA's & ribosomes, force the bacterium to produce hundreds of new viral coats & hundreds of copies of viral DNA.

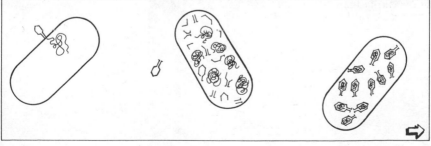

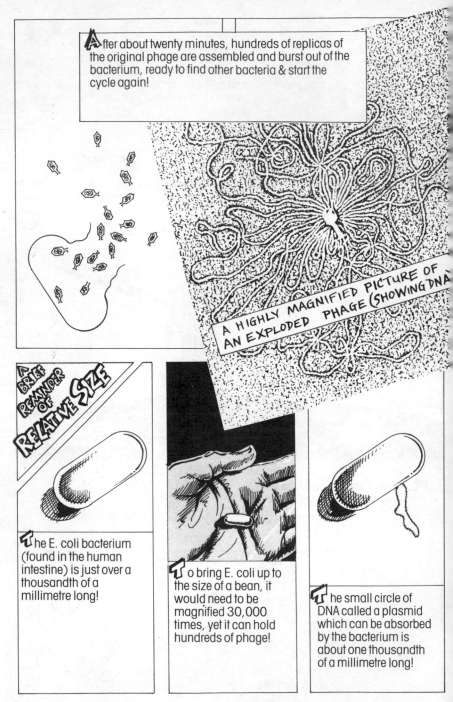

After about twenty minutes, hundreds of replicas of the original phage are assembled and burst out of the bacterium, ready to find other bacteria & start the cycle again!

A HIGHLY MAGNIFIED PICTURE OF AN EXPLODED PHAGE (SHOWING DNA)

A BRIEF REMINDER OF RELATIVE SIZE

The E. coli bacterium (found in the human intestine) is just over a thousandth of a millimetre long!

To bring E. coli up to the size of a bean, it would need to be magnified 30,000 times, yet it can hold hundreds of phage!

The small circle of DNA called a plasmid which can be absorbed by the bacterium is about one thousandth of a millimetre long!

Flabbergasted? Let's further confound you with a glimpse of the complex inner world of the **cell**. A 'typical' cell from a mammal contains enzymes, other proteins, fats, sugars, amino acids, other building blocks & energy – carrying molecules **plus** a yard of double-stranded DNA! The double helix is a little less than 80 billionths of an inch in diameter! A single full twist in the molecule measures just 134 billionths of an inch!

All these complex processes going on in a speck invisible to the eye!! No wonder the beginner is as baffled by the impenetrable 'inner space' of the cell as by the infinite mysteries of the universe!!

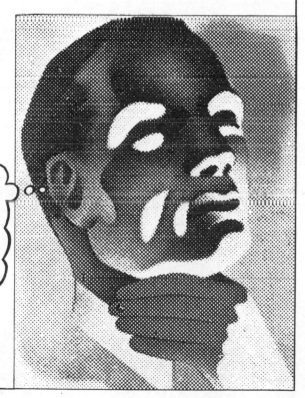

Phew! Having looked at the many different contributors to classical molecular biology, & their discoveries, I think it's time to move into modern genetics, recent research & pointers to the future…

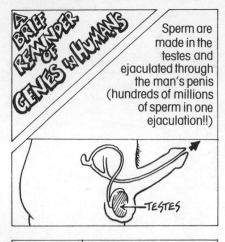

Sperm are made in the testes and ejaculated through the man's penis (hundreds of millions of sperm in one ejaculation!!)

—TESTES

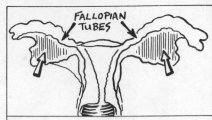

FALLOPIAN TUBES

Eggs are stored in the woman's ovaries & released at the rate of one every four weeks; they lodge in the fallopian tubes awaiting fertilisation by one of the sperm.

A single sperm cell (at the head of the lashing tail) is mainly made up of **nucleus**.

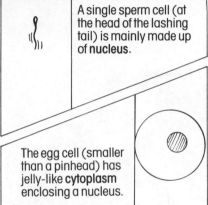

The egg cell (smaller than a pinhead) has jelly-like **cytoplasm** enclosing a nucleus.

Inside the nucleus is a darker-staining material known as **chromatin** made up of fine tangled threads.

Now, when a cell is about to divide the chromatin contracts into groups called chromosomes. A human has 46 chromosomes.

The sperm nucleus has 23 chromosomes (as has the egg nucleus). When sperm fertilises egg the nuclei fuse and the cell has 46 chromosomes once again.

The chromosomes group into 23 **pairs** each with one member from sperm & the other from egg.

To bring one of the larger paired chromosomes up to a centimetre in length it needs to be magnified about 2 thousand times! Chromosomes are composed of the tiny threads of **DNA**.
(MORE ON CHROMOSOMES ON PAGE 145)

With the genetic code cracked, and the outlines of genetic regulation firmly established for bacteria and their viruses, scientists began to confront the awesome problem of gene structure and regulation in higher eukaryotes (those having cells **with** nuclei) including man.

Some dismissed the problem altogether saying:

> The findings are likely to be mere recapitulations of the rules already discovered in bacteria.

Others argued that:

> Fundamental differences between bacteria (which are prokaryotes — cells without nuclei) prevent them from using the very same mechanisms for gene regulation & expression as animal & plants cells (which are eukaryotes & have nuclei).

In **eukaryotes**, because the DNA is contained in the nucleus, it is in a compartment isolated from the translation machinery, which resides in the cytoplasm.

Bacteria, because they lack nuclei, carry out transcription and translation side by side.

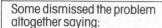

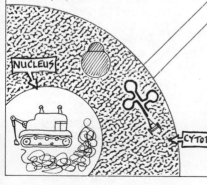

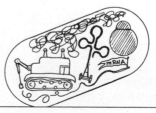

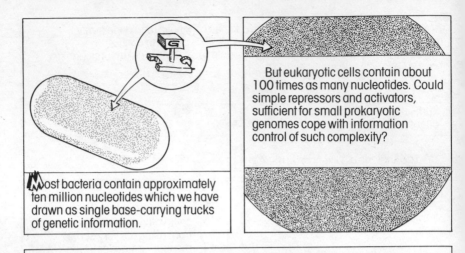

Most bacteria contain approximately ten million nucleotides which we have drawn as single base-carrying trucks of genetic information.

But eukaryotic cells contain about 100 times as many nucleotides. Could simple repressors and activators, sufficient for small prokaryotic genomes cope with information control of such complexity?

Genetics would be one way to study animal cell gene expression, but genetic studies of eukaryotic gene expression were cumbersome. Some of these problems were solved, however, by animal cell tissue culture. In tissue culture, cells are isolated from an animal, and are propagated apart from the animal, in the laboratory, in specially concocted culture media. Unlike bacteria, animal cells are not equipped for infinite division, for in normal circumstances they perish with the death of the individual. After much trial and error, "immortalized" cells, capable of indefinite growth in the laboratory were established.

One human cell line, HeLa, has been propagated in the laboratory over forty years. Cultured cells provide the convenience of bacteria, and permit experiments impossible with the whole animal.

With tissue cultured cells, a form of sexuality may be achieved through "cell fusion".

If different cultured cell types are mingled together in the presence of certain viruses or chemical agents, they clump and merge their exterior membranes. The nuclei of the clumped cells now occupy a single fused cell.

During cell division, the nuclear membranes of the original cells disintegrate, and the sets of chromosones comingle. Cell types from the different species, such as human and mouse cells, can be conveniently fused generating a hybrid man-mouse cell line.

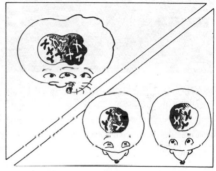

Superficially, cell fusion resembles the comingling of sperm and egg chromosomes after fertilization.

Cells with both human and mouse chromosomes may be obtained in the lab, although they never develop into a hybrid multi-cellular organism such as a mouse-man!

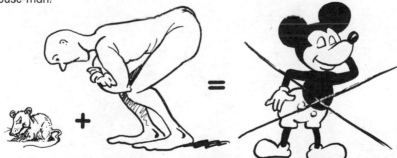

Unfortunately, cells in tissue culture often lose specialized differentiated properties, and their suitability as models for differentiated gene expression is in question.

The problems of **differentiation** raise another major difference between animal cells and bacteria.

A soil bacterium leads a lonely and difficult existence.

It must, from time to time, adapt to change in the nutrients it receives from its surroundings.

A bacterium can undergo many cell divisions without altering the repertoire of gene expression responses it can muster.

Animal cells reside (in general) within the organism, bathed in an unchanging environment of body fluids and tissues. They need not respond to dramatic changes in the environment. Their gene expression programs are geared to doing specific jobs in the organism (differentiation).

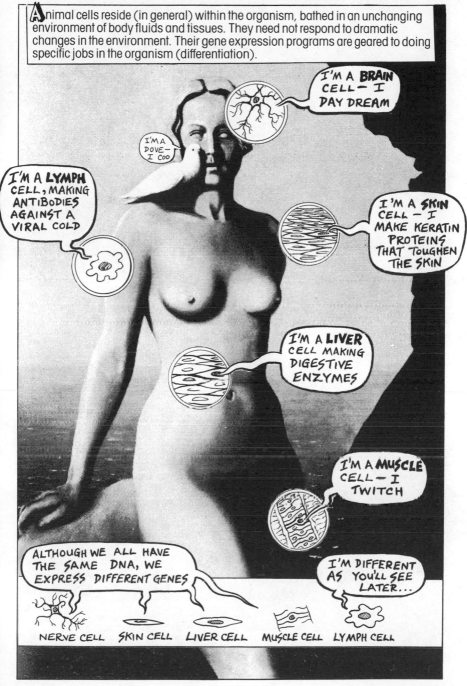

I'M A **BRAIN** CELL — I DAY DREAM

I'M A DOVE — I COO

I'M A **LYMPH** CELL, MAKING ANTIBODIES AGAINST A VIRAL COLD

I'M A **SKIN** CELL — I MAKE KERATIN PROTEINS THAT TOUGHEN THE SKIN

I'M A **LIVER** CELL MAKING DIGESTIVE ENZYMES

I'M A **MUSCLE** CELL — I TWITCH

ALTHOUGH WE ALL HAVE THE SAME DNA, WE EXPRESS DIFFERENT GENES

I'M DIFFERENT AS YOU'LL SEE LATER...

NERVE CELL   SKIN CELL   LIVER CELL   MUSCLE CELL   LYMPH CELL

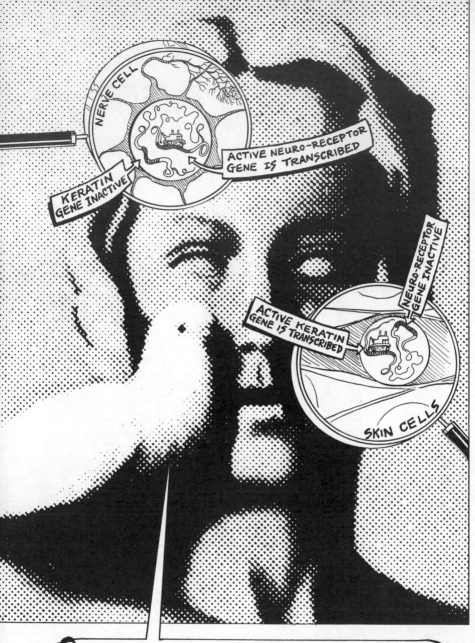

NERVE CELL

ACTIVE NEURO-RECEPTOR GENE IS TRANSCRIBED

KERATIN GENE INACTIVE

NEURO-RECEPTOR GENE INACTIVE

ACTIVE KERATIN GENE IS TRANSCRIBED

SKIN CELLS

During the descent of the differentiated cells from the fertilized egg the very same complement of DNA is retained in each cell (with few exceptions). Differentiation does not result from shedding unwanted genes.

hus, both bacteria and differentiated cells maintain a constant DNA content during gene regulation.

But a bacterium may switch a gene on and off virtually an infinite number of times.

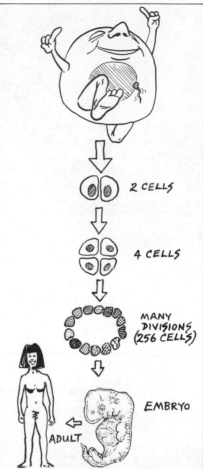

2 CELLS

4 CELLS

MANY DIVISIONS (256 CELLS)

EMBRYO

ADULT

In contrast, a differentiated cell generally does not change its expression to become another type (that is: brain cells cannot become liver cells or vice versa). If it does change the result is often only partial, and can be accompanied by unregulated growth resulting in a tumor.

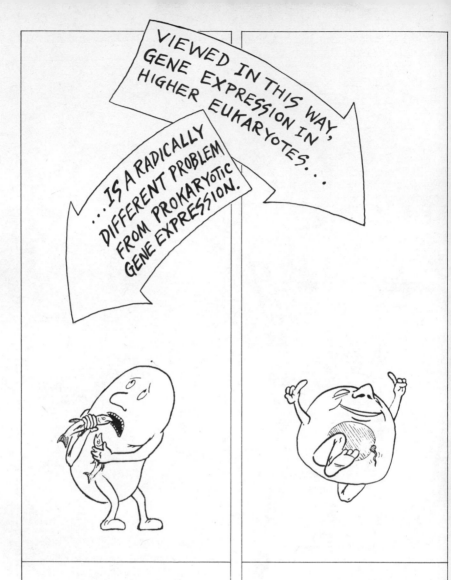

The bacterium copes in a solitary and repetitive manner with fluctuations in its environment, (as the bacterium in the PaJaMo experiment reacts to the absence or presence of sugar).

The eukaryotic fertilized egg throws caution to the wind. It gives rise to a wide range of differentiated cells that form the organism and eventually die. Unlike the bacterium, it can employ regulation mechanisms which are irreversible.

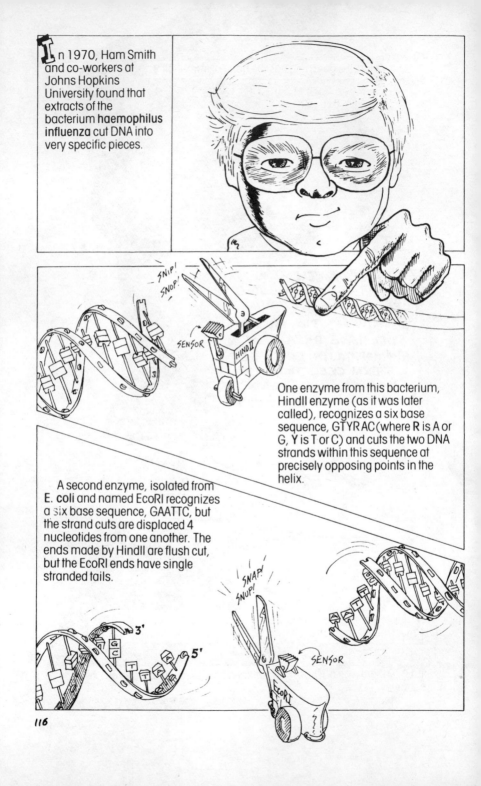

In 1970, Ham Smith and co-workers at Johns Hopkins University found that extracts of the bacterium **haemophilus influenza** cut DNA into very specific pieces.

One enzyme from this bacterium, HindII enzyme (as it was later called), recognizes a six base sequence, GTYRAC (where **R** is A or G, **Y** is T or C) and cuts the two DNA strands within this sequence at precisely opposing points in the helix.

A second enzyme, isolated from E. **coli** and named EcoRI recognizes a six base sequence, GAATTC, but the strand cuts are displaced 4 nucleotides from one another. The ends made by HindII are flush cut, but the EcoRI ends have single stranded tails.

SNIP!
SNOP!
SENSOR
HINDII

SNAP!
SNUP!
SENSOR
EcoRI

3'
G
C
T
5'

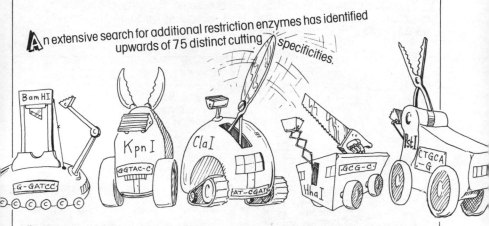

An extensive search for additional restriction enzymes has identified upwards of 75 distinct cutting specificities.

The enzymes are present in a wide range of single-celled organisms: in many bacteria, and in yeast as well. Restriction enzymes provide an invaluable tool for dissecting complex DNA genomes at specific points.

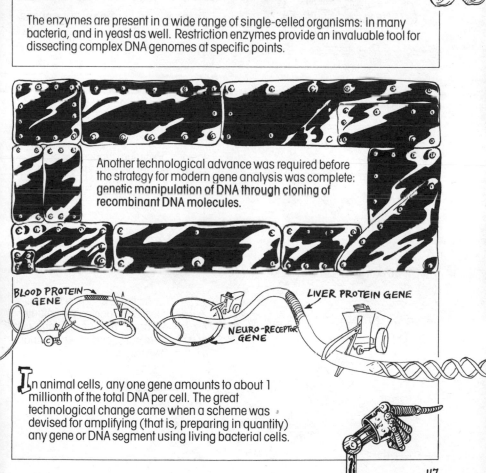

Another technological advance was required before the strategy for modern gene analysis was complete: genetic manipulation of DNA through cloning of recombinant DNA molecules.

BLOOD PROTEIN GENE

LIVER PROTEIN GENE

NEURO-RECEPTOR GENE

In animal cells, any one gene amounts to about 1 millionth of the total DNA per cell. The great technological change came when a scheme was devised for amplifying (that is, preparing in quantity) any gene or DNA segment using living bacterial cells.

We now know how to dissect a gene. But how can we make use of a natural replicating system to amplify a gene – that is, increase the amount of a specific gene by making many identical copies of it?

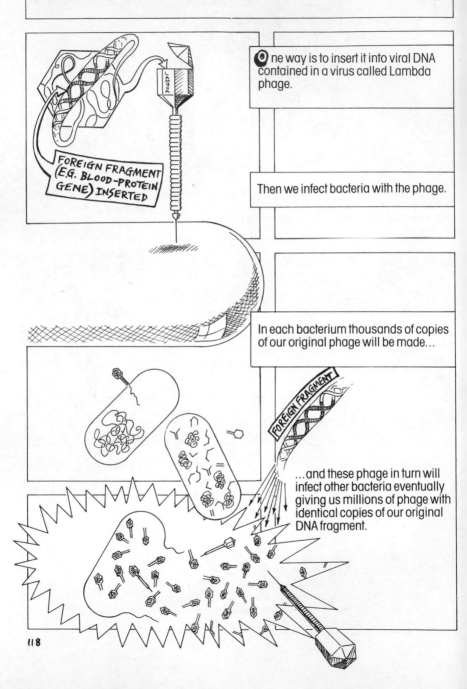

One way is to insert it into viral DNA contained in a virus called Lambda phage.

FOREIGN FRAGMENT (E.G. BLOOD-PROTEIN GENE) INSERTED

Then we infect bacteria with the phage.

In each bacterium thousands of copies of our original phage will be made...

FOREIGN FRAGMENT

...and these phage in turn will infect other bacteria eventually giving us millions of phage with identical copies of our original DNA fragment.

WE CAN INSERT OUR DNA FRAGMENT INTO A PLASMID (A SMALL CIRCULAR PIECE OF DNA)...

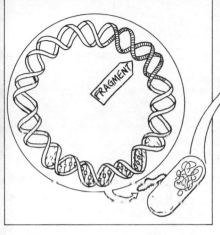

FRAGMENT

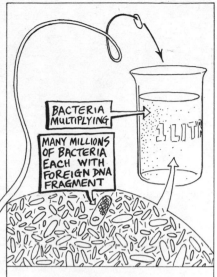

BACTERIA MULTIPLYING

MANY MILLIONS OF BACTERIA EACH WITH FOREIGN DNA FRAGMENT

1 LITR

...the plasmid is taken up by a bacterium & replicates alongside the bacterial DNA

While phage rely on special structures to enter cells, plasmids enter by the inefficient and arduous route of traversing the bacterial membrane as a naked DNA molecule. This closely resembles the entry of transforming principle into bacteria in the Griffith-Avery experiments.

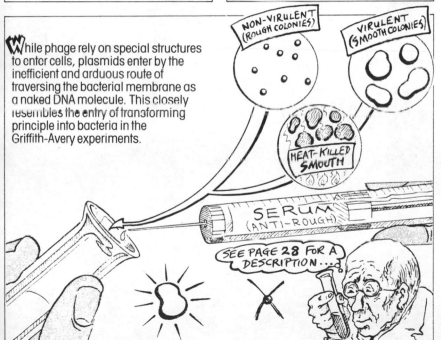

NON-VIRULENT (ROUGH COLONIES)

VIRULENT (SMOOTH COLONIES)

HEAT-KILLED SMOOTH

SERUM (ANTI-ROUGH)

SEE PAGE 28 FOR A DESCRIPTION...

How do we join our foreign gene to the plasmid DNA?

Recall that the EcoRI enzyme makes a staggered break in the double helix.

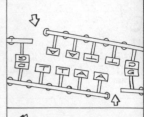

Short, single-stranded tails, four bases long, remain at the cut ends. EcoRI always leaves the tails: AATT.

When a circular DNA molecule (such as a plasmid DNA) which has a single cleavage site is treated with such an enzyme, the circle opens out!

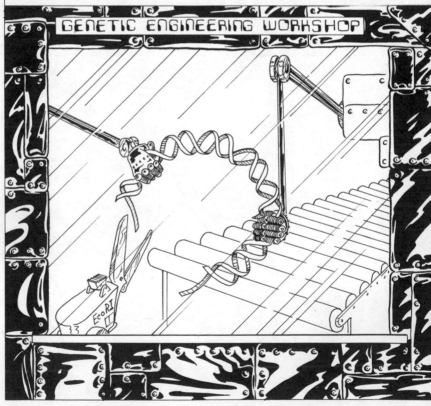

Under certain conditions, the circle can reclose when the single stranded tails bind together again. The base pairing of one tail to the other provides the binding force. The tails are "sticky". Tails of plasmid will stick to the tails of a foreign fragment.

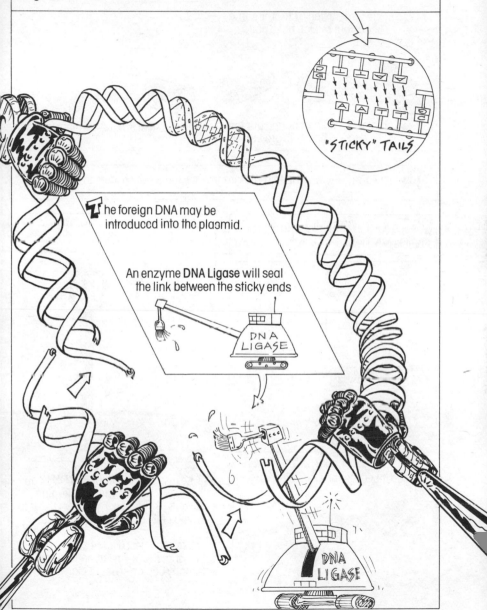

"STICKY" TAILS

The foreign DNA may be introduced into the plasmid.

An enzyme **DNA Ligase** will seal the link between the sticky ends

DNA LIGASE

DNA LIGASE

If we transfer a plasmid vector (several thousand base-pairs)

with foreign DNA thus inserted (recombinant DNA)

to a bacterium having a chromosome (about four million base pairs),

we obtain a strain harbouring our foreign DNA fragment in a form which will be amplified during bacterial growth.

Now there are many ways to join DNA fragments and produce them in quantity.

Phage can be manipulated in a similar way.

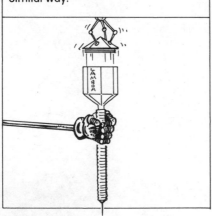

We can insert **foreign DNA** into viral DNA.

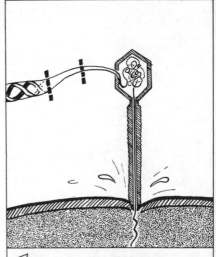

The resulting recombinant DNA can then be incubated with "**packaging extracts**" that reassemble our DNA into biologically active viral structures!

These can infect a large culture of bacteria, make vast quantities of virus, & many copies of our foreign fragment.

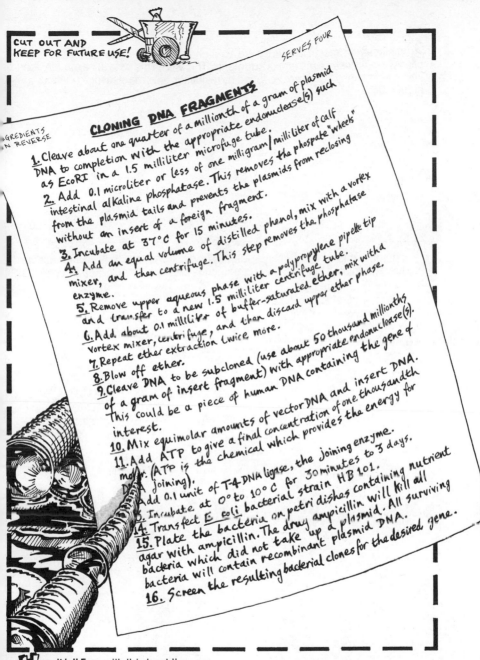

CLONING DNA FRAGMENTS

INGREDIENTS
IN REVERSE

1. Cleave about one quarter of a millionth of a gram of plasmid DNA to completion with the appropriate endonuclease(s) such as EcoRI in a 1.5 milliliter microfuge tube.

2. Add 0.1 microliter or less of one milligram/milliliter of calf intestinal alkaline phosphatase. This removes the phosphate "wheels" from the plasmid tails and prevents the plasmids from reclosing without an insert of a foreign fragment.

3. Incubate at 37°C for 15 minutes.

4. Add an equal volume of distilled phenol, mix with a vortex mixer, and then centrifuge. This step removes the phosphatase enzyme.

5. Remove upper aqueous phase with a polypropylene pipette tip and transfer to a new 1.5 milliliter centrifuge tube.

6. Add about 0.1 milliliter of buffer-saturated ether, mix with a vortex mixer, centrifuge, and then discard upper ether phase.

7. Repeat ether extraction twice more.

8. Blow off ether.

9. Cleave DNA to be subcloned (use about 50 thousand millionths of a gram of insert fragment) with appropriate endonuclease(s). This could be a piece of human DNA containing the gene of interest.

10. Mix equimolar amounts of vector DNA and insert DNA.

11. Add ATP to give a final concentration of one thousandth molar. (ATP is the chemical which provides the energy for DNA joining).

12. Add 0.1 unit of T-4 DNA ligase, the joining enzyme.

13. Incubate at 0° to 10°C for 30 minutes to 3 days.

14. Transfect E. coli bacterial strain HB 101.

15. Plate the bacteria on petri dishes containing nutrient agar with ampicillin. The drug ampicillin will kill all bacteria which did not take up a plasmid. All surviving bacteria will contain recombinant plasmid DNA.

16. Screen the resulting bacterial clones for the desired gene.

Here it is!! Free with this book!!

Your very own recipe for cloning DNA fragments (with explanatory notes).

Sometimes the **foreign DNA** is a pure, well-characterized fragment. However, often we must insert a **mixture** of fragments. This will be the case, for example, if the fragments for insertion were produced by restriction enzyme cleavage of **whole human DNA.**

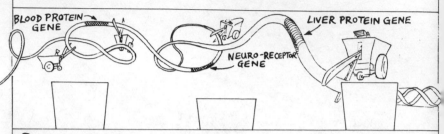

BLOOD PROTEIN GENE

LIVER PROTEIN GENE

NEURO-RECEPTOR GENE

Our purpose might be to amplify & identify a specific human gene. In this case we will have to isolate it from the tens of thousands of other human genes present in human DNA.

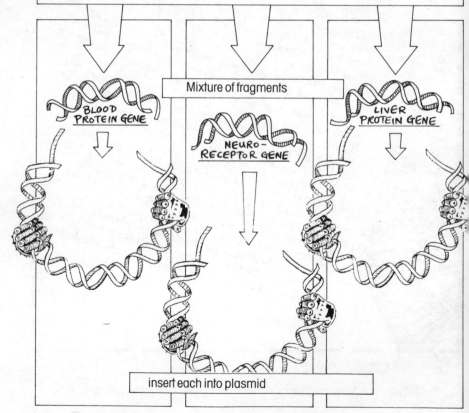

Mixture of fragments

BLOOD PROTEIN GENE

NEURO-RECEPTOR GENE

LIVER PROTEIN GENE

insert each into plasmid

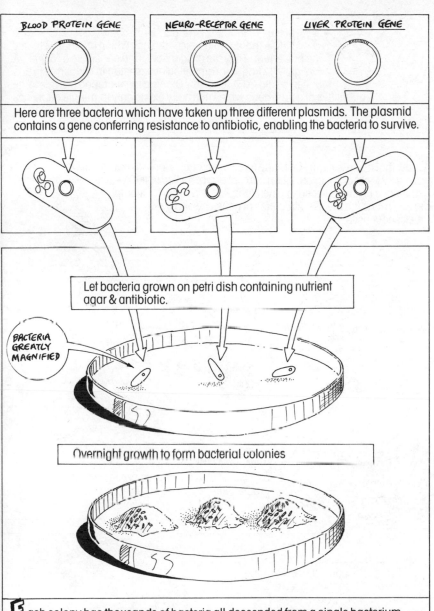

BLOOD PROTEIN GENE

NEURO-RECEPTOR GENE

LIVER PROTEIN GENE

Here are three bacteria which have taken up three different plasmids. The plasmid contains a gene conferring resistance to antibiotic, enabling the bacteria to survive.

Let bacteria grown on petri dish containing nutrient agar & antibiotic.

BACTERIA GREATLY MAGNIFIED

Overnight growth to form bacterial colonies

E ach colony has **thousands** of bacteria all descended from a single bacterium. Each is a **clone**. Each bacterial clone contains one recombinant plasmid type bearing one foreign DNA fragment. We have purified specific foreign fragments from the original mixture.

CLONING DEFINED

All of the plasmids in one bacterial colony descend from a single "parent" plasmid — the one which originally entered the bacterium. They are all identical, and constitute a "clone". A foreign DNA fragment amplified by insertion in a plasmid in this manner is said to be "cloned". Cloning therefore can provide large quantities of a pure gene which normally exists only in minute quantities in the cell.

Now that any gene may be made plentiful through cloning, how shall we study it? Most of the information content of a gene lies in the precise sequence of the nucleotides. Therefore, it was natural to attempt to determine the precise structures of genes by analyzing their sequences.

YAY!

WELCOME BACK, FRED!

WORKERS' GENETIC SEQUENCER 1ST CLASS

The master of sequencing, **Fred Sanger**, champion of proteins with his analysis of insulin had, by 1965, graduated to sequencing RNA.

With the challenge of DNA looming before him, Sanger devised a technique for DNA sequencing which used cloning technology and DNA synthesis enzymology.

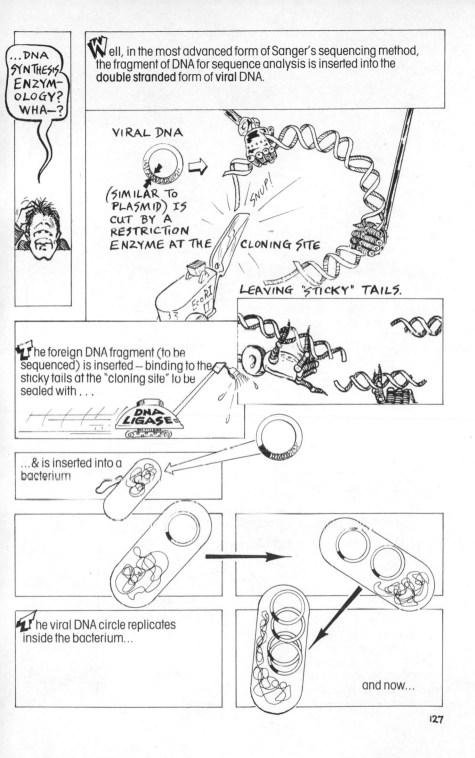

...DNA SYNTHESIS ENZYMOLOGY? WHA—?

Well, in the most advanced form of Sanger's sequencing method, the fragment of DNA for sequence analysis is inserted into the **double stranded** form of **viral DNA**.

VIRAL DNA

(SIMILAR TO PLASMID) IS CUT BY A RESTRICTION ENZYME AT THE CLONING SITE

SNIP!

EcoRI

LEAVING "STICKY" TAILS.

The foreign DNA fragment (to be sequenced) is inserted — binding to the sticky tails at the "cloning site" to be sealed with . . .

DNA LIGASE

...& is inserted into a bacterium

The viral DNA circle replicates inside the bacterium...

and now...

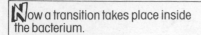

Now a transition takes place inside the bacterium.

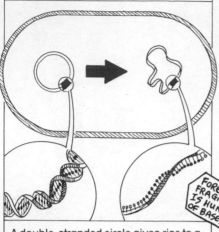

A double-stranded circle gives rise to a single-stranded circle by a complex process!

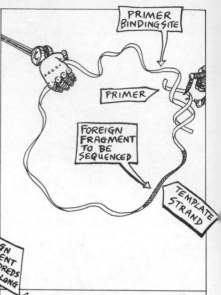

PRIMER BINDING SITE

PRIMER

FOREIGN FRAGMENT TO BE SEQUENCED

TEMPLATE STRAND

FOREIGN FRAGMENT IS HUNDREDS OF BASES LONG

The single-stranded DNA is purified from the bacterium & a short piece of DNA called a **Primer** (15 bases long), whose sequence is exactly complementary to the single strand next to the cloning site, is added.

DNA POLYMERASE

Here's **DNA Polymerase** with its driver. It's specially built to copy single-stranded DNA, but needs a **double-stranded start point**. This is provided by the **Primer**.

Now remember that DNA polymerase is surrounded by many normal free-floating DNA base trucks.

Sanger provided these trucks but also a very few "chain terminating inhibitor trucks" (dideoxy-triphosphates).

STOP

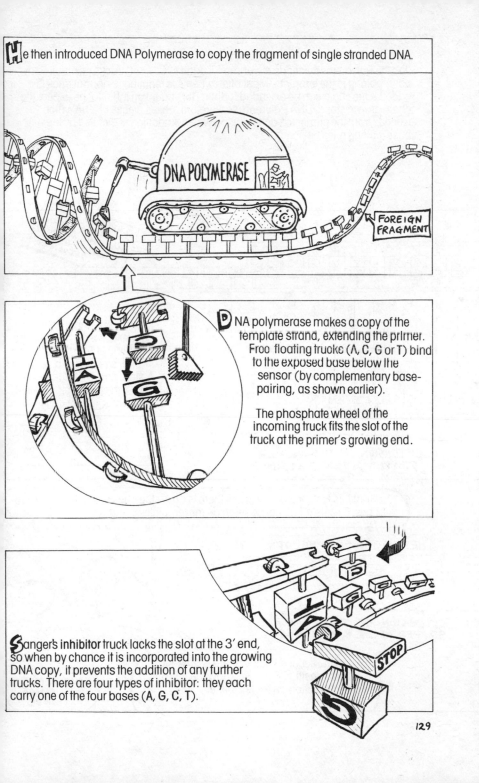

He then introduced DNA Polymerase to copy the fragment of single stranded DNA.

DNA POLYMERASE

FOREIGN FRAGMENT

DNA polymerase makes a copy of the template strand, extending the primer. Free floating trucks (A, C, G or T) bind to the exposed base below the sensor (by complementary base-pairing, as shown earlier).

The phosphate wheel of the incoming truck fits the slot of the truck at the primer's growing end.

Sanger's **inhibitor** truck lacks the slot at the 3' end, so when by chance it is incorporated into the growing DNA copy, it prevents the addition of any further trucks. There are four types of inhibitor: they each carry one of the four bases (A, G, C, T).

STOP

As the Polymerase extends the primer, it adds A, G, T or C trucks as required by base-pairing to the template. Most often when A is required, a normal truck is added & the chain can be extended further. But should an inhibitor be joined, the chain growth stops. DNA Polymerase therefore makes pieces of DNA which extends from the primer to each position where A addition is dictated by the base-pairing rules.

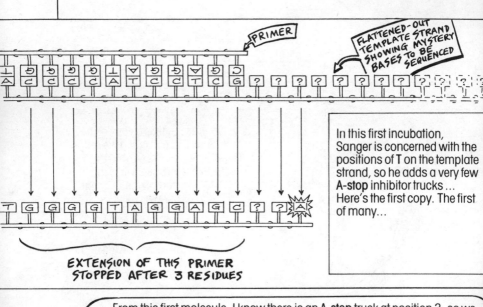

PRIMER

FLATTENED-OUT TEMPLATE STRAND SHOWING MYSTERY BASES TO BE SEQUENCED

EXTENSION OF THIS PRIMER STOPPED AFTER 3 RESIDUES

In this first incubation, Sanger is concerned with the positions of **T** on the template strand, so he adds a very few **A-stop** inhibitor trucks … Here's the first copy. The first of many…

From this first molecule, I know there is an **A-stop** truck at position 3, so we know there is a T at that position on the template.

HERE'S THE SECOND COPY

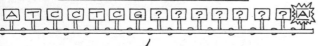

EXTENSION OF THIS PRIMER STOPPED AFTER 8 RESIDUES

From the second, I know there is an A-stop truck at position 8, so we know there is a T at that position on the template.

HELL! THIS GUY SANGER MUST HAVE **SOME** PATIENCE!

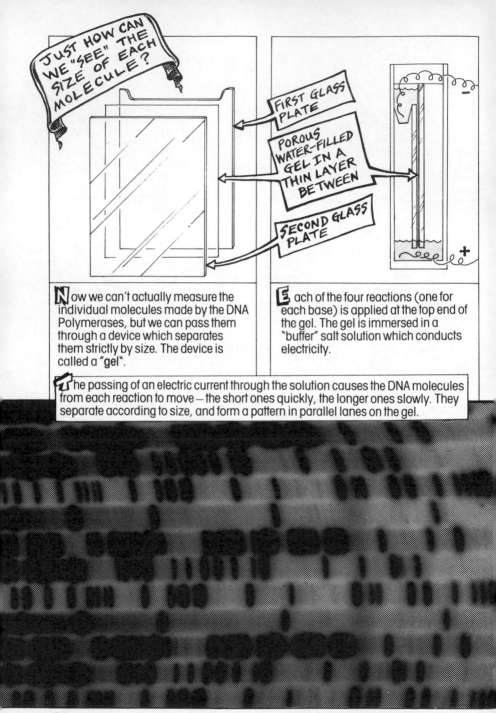

JUST HOW CAN WE "SEE" THE SIZE OF EACH MOLECULE?

FIRST GLASS PLATE

POROUS WATER-FILLED GEL IN A THIN LAYER BETWEEN

SECOND GLASS PLATE

Now we can't actually measure the individual molecules made by the DNA Polymerases, but we can pass them through a device which separates them strictly by size. The device is called a "gel".

Each of the four reactions (one for each base) is applied at the top end of the gel. The gel is immersed in a "buffer" salt solution which conducts electricity.

The passing of an electric current through the solution causes the DNA molecules from each reaction to move — the short ones quickly, the longer ones slowly. They separate according to size, and form a pattern in parallel lanes on the gel.

132

…the molecular "rulers" make a pattern on an X-ray film and allow me to read the structure!

The genius of Sanger's method was that the pattern could be interpreted directly to give the complete structure — 400 bases in a single experiment — of the gene.

Meanwhile in the other Cambridge — at Harvard in Massachusetts — **Walter Gilbert** muses…

Very clever, Fred! and now Allan Maxam & I have developed a different way of making patterns of DNA structure by breaking the molecules down by four separate chemical degradations — one for each type of base!

and it's just as easy to interpret!

OF COURSE, MY CROWNING SUCCESS WAS THE UNRAVELLING OF THE STRUCTURE OF LAMDA-PHAGE DNA - NEARLY 50,000 BASES LONG!

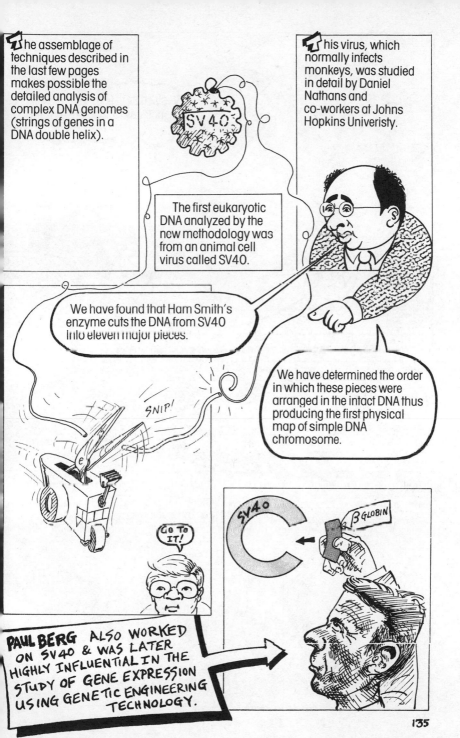

The assemblage of techniques described in the last few pages makes possible the detailed analysis of complex DNA genomes (strings of genes in a DNA double helix).

This virus, which normally infects monkeys, was studied in detail by Daniel Nathans and co-workers at Johns Hopkins Univeristy.

SV40

The first eukaryotic DNA analyzed by the new methodology was from an animal cell virus called SV40.

We have found that Ham Smith's enzyme cuts the DNA from SV40 into eleven major pieces.

We have determined the order in which these pieces were arranged in the intact DNA thus producing the first physical map of simple DNA chromosome.

SNIP!

GO TO IT!

SV40

β GLOBIN

PAUL BERG ALSO WORKED ON SV40 & WAS LATER HIGHLY INFLUENTIAL IN THE STUDY OF GENE EXPRESSION USING GENETIC ENGINEERING TECHNOLOGY.

135

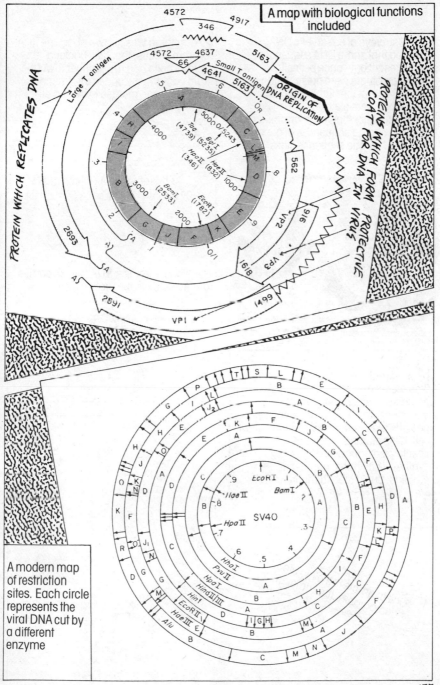

A map with biological functions included

PROTEIN WHICH REPLICATES DNA

PROTEINS WHICH FORM PROTECTIVE COAT FOR DNA IN VIRUS

Large T antigen

Small T antigen

ORIGIN OF DNA REPLICATION

VP2

VP3

VP1

A modern map of restriction sites. Each circle represents the viral DNA cut by a different enzyme

SV40

EcoRI
BamI
Hae II
Hpa II
HhaI
Pvu II
Hpa I
Hind II/III
Hinf
EcoRII
Hae III
Alu

Early surveys of eukaryotic DNA revealed that some DNA sequences were present only once per cell. However, others were present many times. Some of the highly repeated sequences were likely to have a **structural** rather than an **informational** role. The coding sequences of genes fell in the "**unique sequence class**".

For animal cell gene expression, it was messenger RNA, and its precursors which were most important. And here the difference with bacteria was dramatic. The first animal cell RNAs studied were from specialized animal cells that synthesize a limited number of proteins, but in great quantity.

Something was known about the RNA of animal cells, too. Like bacteria, animal cells had: **messenger RNA** (the train carrying the coded information from RNA polymerase to the protein factory),

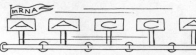

— **ribosomal RNA** (which makes up the protein factory),

and — **transfer RNA** (which carries amino acids to the protein factory).

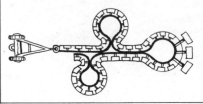

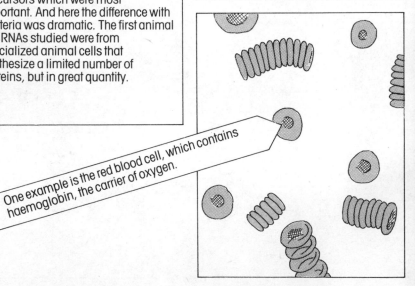

One example is the red blood cell, which contains haemoglobin, the carrier of oxygen.

The protein component, globin, is translated from an mRNA which is plentiful in red blood cells and easily purified.

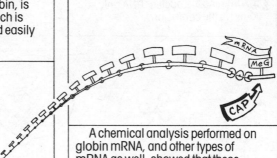

A chemical analysis performed on globin mRNA, and other types of mRNA as well, showed that these molecules had unexpected modifications not found in bacterial mRNA.

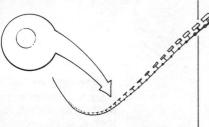

At the 5' end they had an unprecedented "inverted G" residue, called a **cap**. And at the 3' end they had a long string of A's, up to 200 in number, called "poly A".

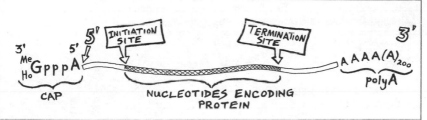

No equivalent to poly A or the cap are in the DNA, and these are added to the mRNA by special mechanisms after RNA transcription!

Messenger RNA was made from RNA in the nucleus (nuclear RNA), but how?

In bacteria, the RNA transcript, as originally copied from DNA, is the mRNA.

But in eukaryotes scientists asked **IS THERE A "PRE-MESSENGER"?**

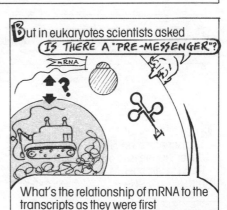

What's the relationship of mRNA to the transcripts as they were first synthesized in the nucleus?

E xamining the nuclear RNA only added to the confusion.

First there was a lot of it!

In fact about nine parts of nuclear RNA were made by the mobile scanner for every part that actually reached the cytoplasm as mRNA...

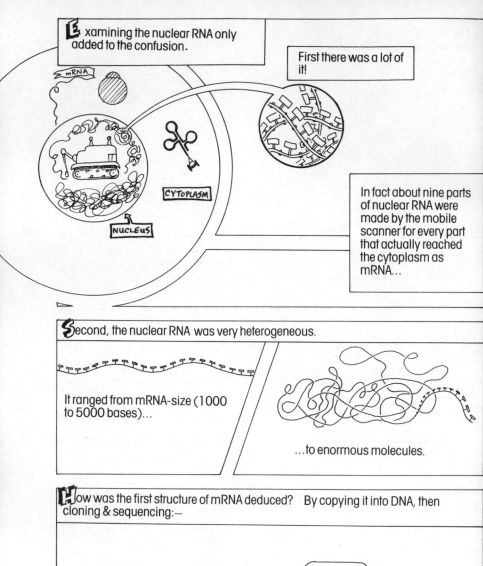

mRNA

CYTOPLASM

NUCLEUS

S econd, the nuclear RNA was very heterogeneous.

It ranged from mRNA-size (1000 to 5000 bases)...

...to enormous molecules.

H ow was the first structure of mRNA deduced? By copying it into DNA, then cloning & sequencing:—

H ere's **Reverse Transcriptase** and its driver . Its job is to copy RNA (in this case mRNA) into DNA.

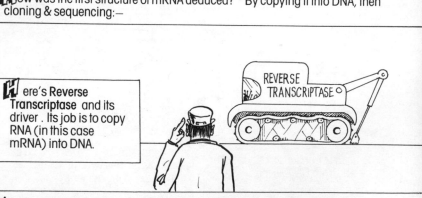

REVERSE TRANSCRIPTASE

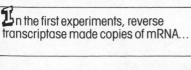
In the first experiments, reverse transcriptase made copies of mRNA...

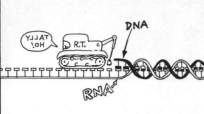

This copy of mRNA was converted to a DNA double helix...

Here's how...

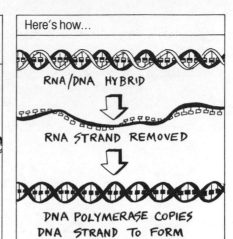

RNA/DNA HYBRID

RNA STRAND REMOVED

DNA POLYMERASE COPIES
DNA STRAND TO FORM
A NEW DNA DOUBLE HELIX

The resulting double-stranded DNA is cloned. By sequencing these clones, the complete structure of globin mRNA was soon known.

The next task was to examine the gene from which the message was copied.

This was achieved by a novel application of the hybridization technique in which mRNA was forced to base-pair with its template DNA.

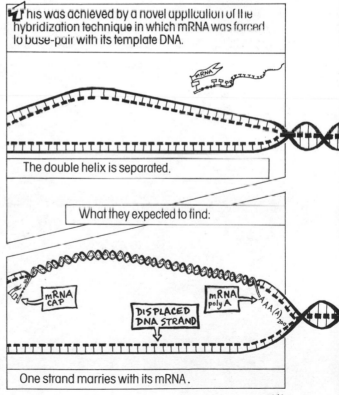

The double helix is separated.

What they expected to find:

mRNA CAP

DISPLACED DNA STRAND

mRNA poly A

AAA(A) 200

One strand marries with its mRNA.

141

This hybrid of messenger RNA & a template strand of genomic DNA was viewed under an electron microscope ... instead of seeing the thick double-stranded hybrid of RNA & DNA — called a **heteroduplex** — with thinner single strands of DNA extending beyond the mRNA, the results were amazing & completely unexpected!

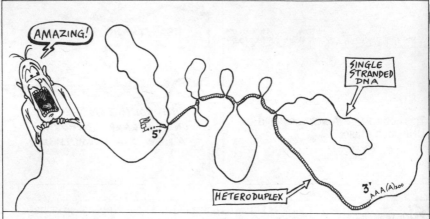

Researchers saw a much more complex pattern: double strands were followed by DNA single strands — but further along more double and single strands appeared!

The DNA of the genes seemed to be in sections spread through the genome with bits of extraneous DNA in between. Let's look at the β Globin gene:

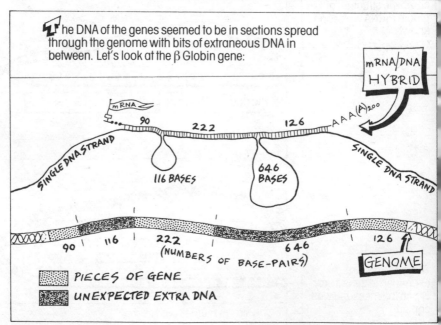

The discovery was made with mRNA from Adenovirus, & it was soon confirmed with cellular genes. The globin gene from the red blood cell was in 3 pieces, with two extraneous bits unexplainably inside the gene!

This complexity posed a difficult problem for gene researchers.

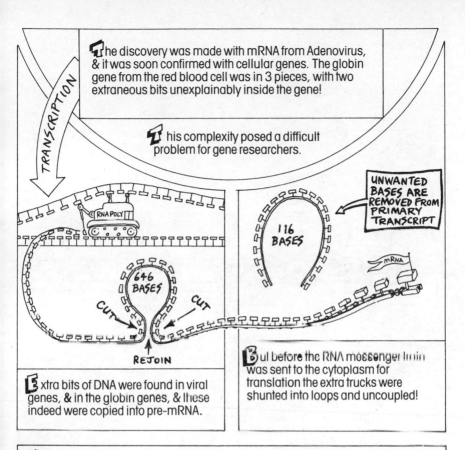

TRANSCRIPTION

RNA POLY

646 BASES

CUT

CUT

REJOIN

Extra bits of DNA were found in viral genes, & in the globin genes, & these indeed were copied into pre-mRNA.

UNWANTED BASES ARE REMOVED FROM PRIMARY TRANSCRIPT

116 BASES

mRNA

But before the RNA messenger train was sent to the cytoplasm for translation the extra trucks were shunted into loops and uncoupled!

We can imagine the primary transcript as a manuscript with sections of gibberish in the text. Garbled parts have to be removed by an editor before the book can be published.

The reaction which ~~grn splk hog uni spoilg~~ch removed the extra sequences (called **introns**) and re~~gazo oglytr ynpcehlet~~ joined the mRNA seque~~premurgle astand tweeple~~ ~~aras nisma~~tnces (called exons) is called RNA Splicing.

The reaction which removed the extra sequences (called introns) and rejoined the mRNA sequences (called exons) is called **RNA Splicing**.

The existence of split genes & **RNA splicing** was completely unexpected! It reaffirmed the difference between eukaryotic & prokaryotic genes…

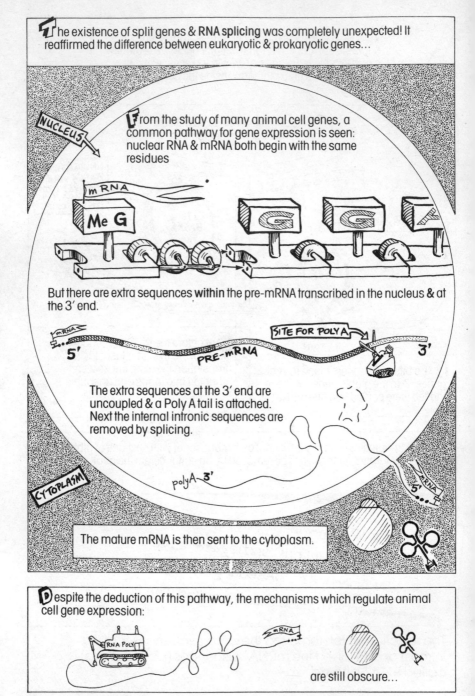

From the study of many animal cell genes, a common pathway for gene expression is seen: nuclear RNA & mRNA both begin with the same residues

mRNA

Me G   G   G   G

NUCLEUS

But there are extra sequences **within** the pre-mRNA transcribed in the nucleus & at the 3′ end.

mRNA
5′   PRE-mRNA   SITE FOR POLY A   3′

The extra sequences at the 3′ end are uncoupled & a Poly A tail is attached. Next the internal intronic sequences are removed by splicing.

polyA 3′

CYTOPLASM

mRNA
5′...

The mature mRNA is then sent to the cytoplasm.

Despite the deduction of this pathway, the mechanisms which regulate animal cell gene expression:

RNA POLY   mRNA

are still obscure…

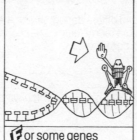

Many genes are transcriptionally controlled.

For some genes repressors or activators which bind to the promoter are likely to control transcription, just as in bacterial genes.

However for many eukaryotic genes, the structure of the **chromatin** (dark-staining material inside the nucleus) may be critical.

Chromatin is a complex of eukaryotic DNA with positively charged proteins called **histones**.

DNA winds twice about the core to form the fundamental subunit of chromatin called the **nucleosome.**

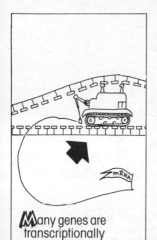

HISTONE

The histones form a nucleosome core.

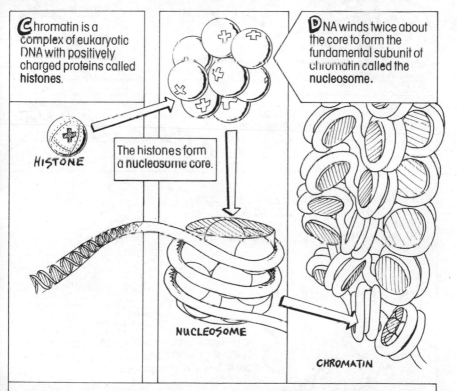

NUCLEOSOME

CHROMATIN

Chromatin consists of many nucleosomes linked by DNA & packaged into more complex but regular fibres.

The structure of the chromatin differs depending on the activity of the gene. Open conformations of chromatin are ready for transcription while inactive chromatin is closed to the surroundings & compact.

A new form of DNA — called **Z DNA** — has been found by Alex Rich at M.I.T....

which winds in a left-handed double helix!

GREAT! JUST WHEN I WAS GETTING THE HANG OF IT

The **A & B forms of DNA** studied by Watson, Crick, Franklin & Wilkins were always right-handed.

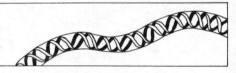

Z-DNA has a left-handed winding in a zig-zag spiral as shown below (as opposed to the smooth spiral of A & B DNA) which removes DNA torsional (twisting) stress likely to be required for transcription...

SNIFF! — (REDUNDANT)

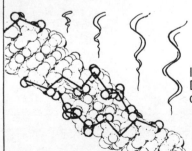

In chromatin, by the twisting of a segment of the DNA into the left-handed Z form, adjacent genes are thought to be regulated.

During development, different globin proteins are expressed: the embryonic, the foetal, and finally the adult globin. The developing organism's requirements for transporting oxygen change as it grows from embryo to foetus to adult. Therefore different forms of the oxygen-carrying globin protein are produced through the successive activation of genes for:

1. **Embryonic**
(up to 12 weeks)

2. **Foetal**
(up to birth)

3. **Adult** globin.
(from birth)

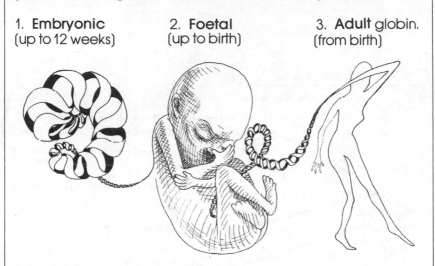

The genes for the different globin types are situated together within a 40,000 base region of the human genome. They form a **gene family**.

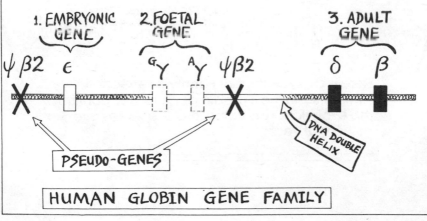

HUMAN GLOBIN GENE FAMILY

Now let's put the human globin genes into context with similar genes in other primates.

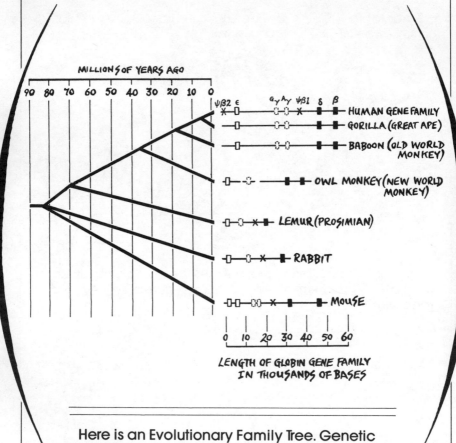

Here is an Evolutionary Family Tree. Genetic similarities in the β globin gene family confirm the evolutionary relationship of species over millions of years.

Although embryonic, foetal and adult genes are different and suited to the specific stages of development, nucleotide sequences show them to be closely related. Perhaps they had a primordial gene as a common ancestor. Duplication of this primordial gene would have allowed the separate copies to evolve to their current structures.

Changes in chromatin structure of this globin cluster during development may switch expression from embryonic to foetal to adult globin. The globin cluster also contains a non-fuctional relic of a globin gene which was inactivated through mutation. This is called a globin pseudo-gene.

GREAT EXPIRING ILLUSTRATORS

No. 98   Borin Van Loon

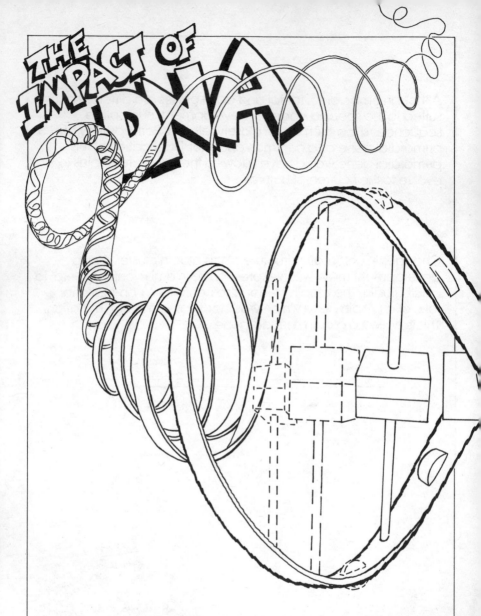

THE IMPACT OF DNA

Cloning not only taught us much about gene structure, it also captured the public imagination – and later its fears. The first public debate over the benefits and hazards of genetic engineering came in 1973 when scientists active in gene research considered the implications of cloning for society.

151

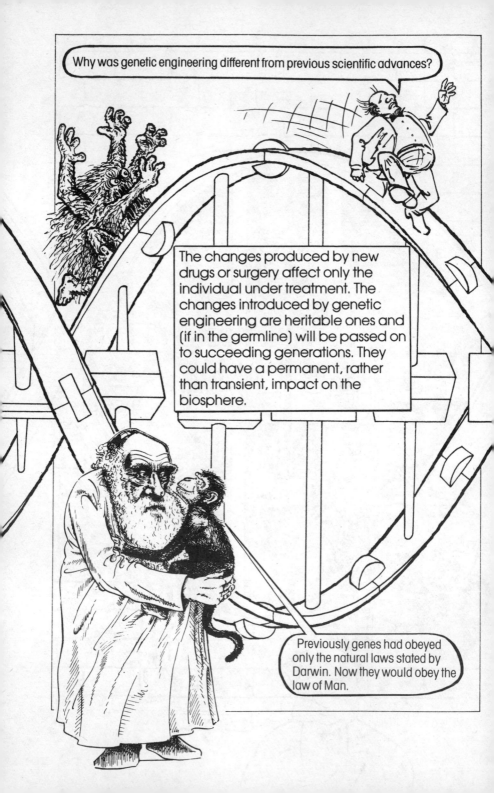

Biologists recognized that the most likely effect of reorganizing DNA would be to make it, simply stated, non-functional. Genes have evolved through eons of selection, mutational change and natural trial.

Transferring human genes into bacteria would therefore probably disrupt their function rather than create a hazard in the biosphere.

However, concern remained that a novel recombinant could have undesirable properties. For example, a bacterium synthesizing insulin in the gut of a human might imbalance the person's sugar metabolism.

With a call for a moratorium in 1974, scientists voluntarily deferred certain classes of cloning experiments. Later strict guidelines for conducting cloning were adopted by the scientific community.

Some groups, such as Science for the People – a New Left coalition of scientists from the Boston-Cambridge area – opposed the guidelines as a breach of the moratorium.

Others argued that experiments should be allowed to continue without bureaucratic regulations at all, be it moratorium or guideline. Some experiments, such as cloning random fragments of primate DNA, required such rigorous containment facilities or enfeebled bacteria hosts that, effectively, they were beyond the means of virtually all laboratories.

As the issues of recombinant DNA were taken up by higher and more bureaucratic agencies, the public press, local government and scientific partisans heated the debate.

In virtually every instance, the discussion was of the perils of recomb-inant DNA.

The Cambridge, Mass. City Council attempted to ban recombinant DNA research from the prestigious Harvard Biological Laboratories.

One critic questioned:

What if an ant crawled from the bio-containment facility with a perilous **E. coli** perched on its thorax?

Opposing scientists grappled in public debate.

Most poignantly, Erwin Chargaff and James Watson who once exchanged scornful and mistrusting glances in the Cavendish Laboratories, assaulted each other's positions on recombinant DNA.

Said Chargaff in 1976:

I don't know whether there could be epidemics, but that possibility we raise in the public's mind and on the part of many countries. This question is itself enough for me to advocate the most strict controls.

Said Watson in 1979:

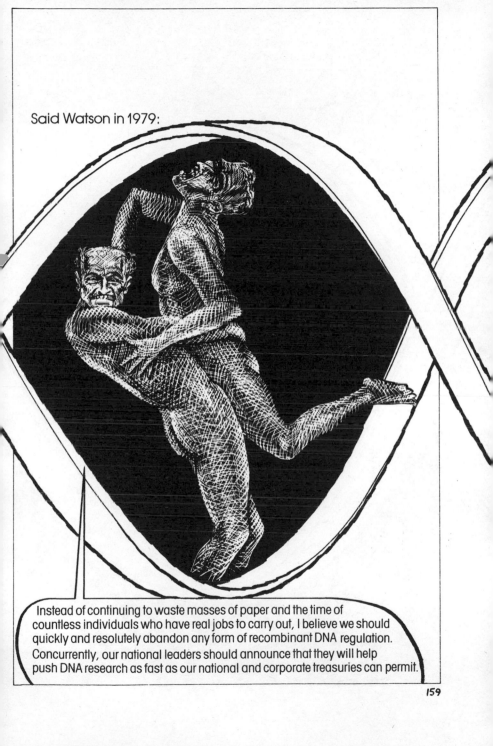

Instead of continuing to waste masses of paper and the time of countless individuals who have real jobs to carry out, I believe we should quickly and resolutely abandon any form of recombinant DNA regulation. Concurrently, our national leaders should announce that they will help push DNA research as fast as our national and corporate treasuries can permit.

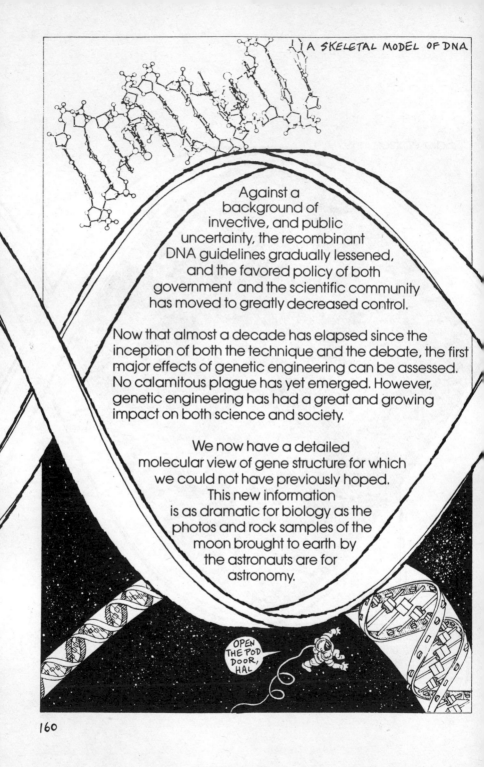

A SKELETAL MODEL OF DNA

Against a background of invective, and public uncertainty, the recombinant DNA guidelines gradually lessened, and the favored policy of both government and the scientific community has moved to greatly decreased control.

Now that almost a decade has elapsed since the inception of both the technique and the debate, the first major effects of genetic engineering can be assessed. No calamitous plague has yet emerged. However, genetic engineering has had a great and growing impact on both science and society.

We now have a detailed molecular view of gene structure for which we could not have previously hoped. This new information is as dramatic for biology as the photos and rock samples of the moon brought to earth by the astronauts are for astronomy.

OPEN THE POD DOOR, HAL

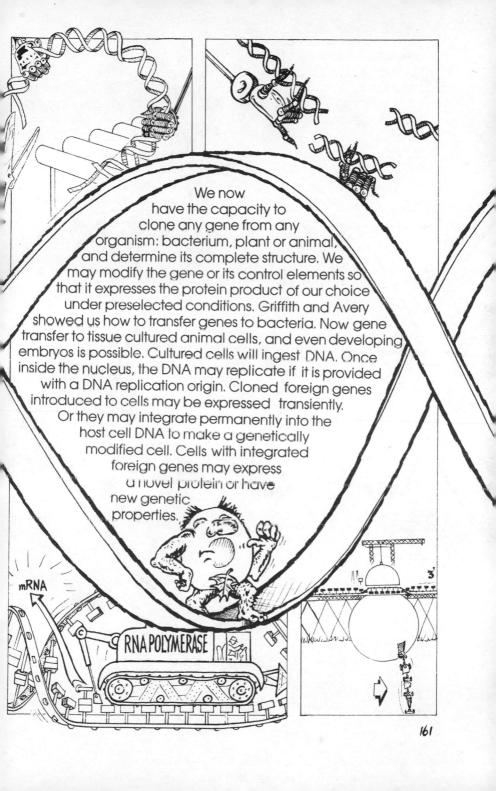

We now have the capacity to clone any gene from any organism: bacterium, plant or animal, and determine its complete structure. We may modify the gene or its control elements so that it expresses the protein product of our choice under preselected conditions. Griffith and Avery showed us how to transfer genes to bacteria. Now gene transfer to tissue cultured animal cells, and even developing embryos is possible. Cultured cells will ingest DNA. Once inside the nucleus, the DNA may replicate if it is provided with a DNA replication origin. Cloned foreign genes introduced to cells may be expressed transiently. Or they may integrate permanently into the host cell DNA to make a genetically modified cell. Cells with integrated foreign genes may express a novel protein or have new genetic properties.

mRNA

RNA POLYMERASE

3'

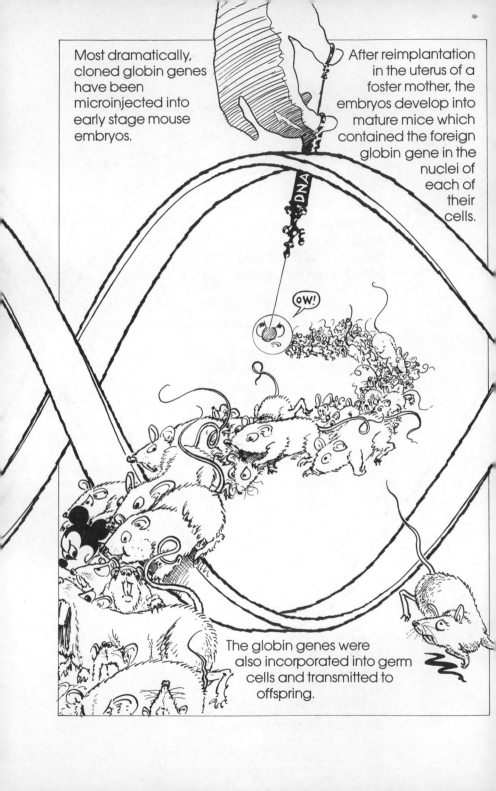

Expression of transferred foreign genes may change the metabolism or growth properties of a cell. In early gene transfer experiments, foreign genes restored the ability of cells to grow despite lethal mutations, or to survive in the prescence of growth inhibitors.

One of the most dramatic breakthroughs has been in the understanding of the causes of certain cancers. Cancer is a group of related diseases in which cells fail to observe the normal signals which regulate cell proliferation. At least some cancers arise from genetic changes which occur in somatic cells. The mutated cells divide when they shouldn't, and form a tumor.

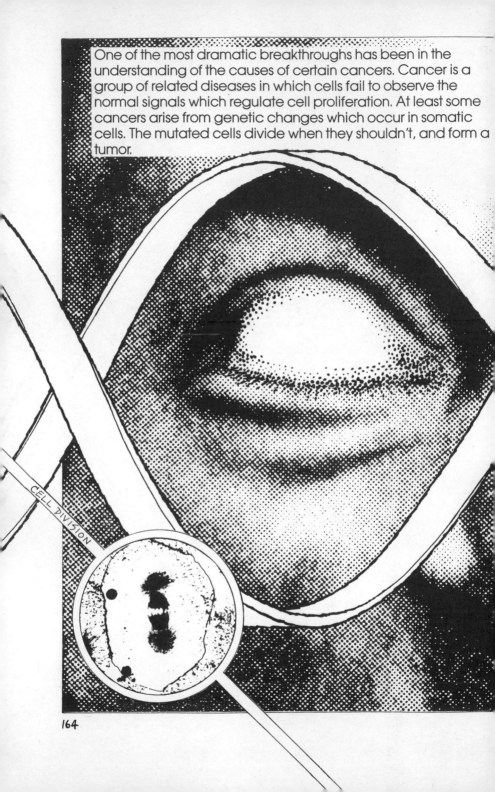

CELL DIVISION

Robert Weinberg of M.I.T., Geoffrey Cooper of Harvard and Michael Wigler of Cold Spring Harbour Labs in New York used gene transfer to identify so called "cancer genes": genes which cause uncontrolled growth of malignant cells.

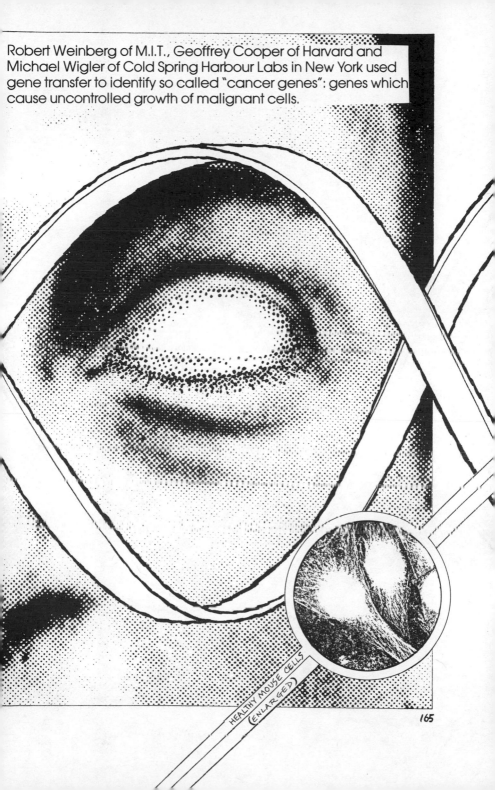

HEALTHY MOUSE CELLS
(ENLARGED)

DNA from human tumor cells was transferred to normal mouse cells in tissue culture. After ingesting the DNA, some of the normal mouse cells commenced uncontrolled growth. The workers hypothesized that expression of a gene present in the human tumor DNA altered the mouse cell growth.

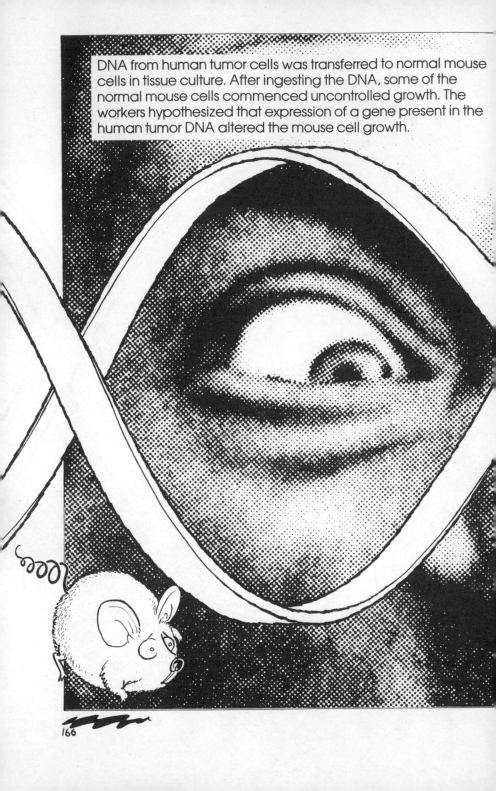

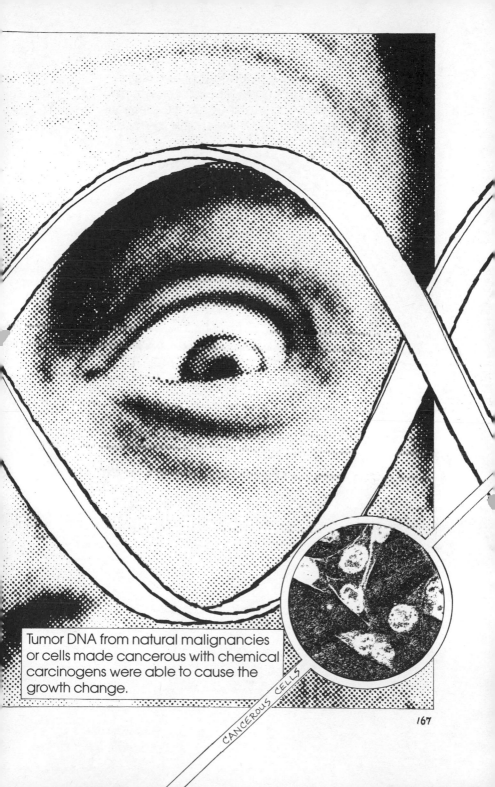

Tumor DNA from natural malignancies or cells made cancerous with chemical carcinogens were able to cause the growth change.

CANCEROUS CELLS

The cancer genes responsible for uncontrolled growth were cloned, and it was soon evident that the cancer genes had nearly identical relatives in normal cells. Apparently normal cellular genes can cause malignant growth after they have undergone subtle alterations which (most likely) alter the mode of their expression, or a critical aspect of the structure of their protein product.

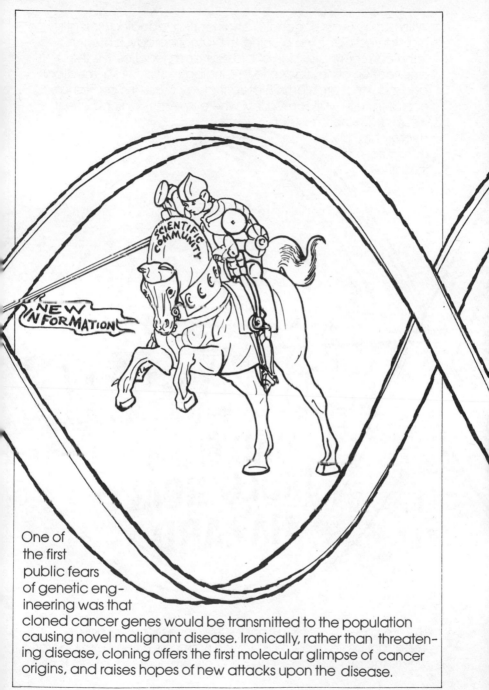

One of
the first
public fears
of genetic eng-
ineering was that
cloned cancer genes would be transmitted to the population
causing novel malignant disease. Ironically, rather than threatening disease, cloning offers the first molecular glimpse of cancer
origins, and raises hopes of new attacks upon the disease.

With public fears of genetic engineering receding, scientists considered applying cloning to biotechnology. Biotechnology is the commercialization of biology and genetics. It is the application of new genetic technology to practical medical and industrial problems. Biotechnology arose in San Francisco, not far from the Silicon Valley, where investors gained great returns from the transistor, micro-chip, and computer industries.

The first projects were to transfer the genes for medically important proteins such as growth hormone or insulin to bacteria where the proteins might be cheaply produced in abundance.

The first biotechnology experiments were conducted in University research labs, on work benches neighboring basic research projects financed by government funds. Hostilities between the basic and applied camps sometimes raged.

DESOXYRIBONUCLEIQUE?

Plin Augé

Yves St. Agnes' dna cream

But soon investors became enchanted with genetic engineering and tens of millions of dollars of venture capital were provided to build laboratory facilities for new biotechnology companies.

The giddy stage of investment passed and many ask how, realistically, biotechnology might profit society. Here are some possiblities. Biotechnology promises to:

1. Provide diagnostic reagents for detecting genetic diseases such as Down's syndrome, sickle cell anemia, or even somatic genetic diseases such as cancer.

2. Produce vaccines against diseases of livestock and (with government approval) humans. Some vaccines, such as for malarial parasites – which cleverly avoid immune detection – might not be feasible through other approaches.

3. Produce hormones, blood clotting factors, insulin or other protein pharmaceuticals such as interferon. In the future new complex and specifically targeted "protein drugs" may be possible with gene cloning.

4. Produce industrial chemicals such as the sweetener fructose. Woodchips might be converted to sugar or to synthetic fuels such as gasohol.

5. Genetically modify plants for the mass production of chemicals and proteins and novel nutrients fueled by cheap photosynthetic energy. Bypass natural barriers for gene transfer and overcome slow breeding times to make plants disease-resistant and viable in soils previously unsuitable for agriculture.

6. Develop new energy sources and animal feed stocks through waste and biomass recycling.

7. Salvage precious metals, develop new bio-mining techniques for the recovery of ore metals. Control pollution.

173

In the nineteenth century the Swedish physicist Gustav Arrhenius suggested the theory of **panspermia**.

"Life originated elsewhere in the universe and has been brought to Earth by microorganisms."

A few years ago, Leslie Orgel and Francis Crick resuscitated the theory:

But what if life did begin on earth. How could it have started? In 1924 the Russian biochemist A. I. Oparin published a monograph (little noticed at the time):

That early atmosphere of the Earth probably contained methane gas, ammonia and water, but **no** oxygen. Ultraviolet light from the sun, electrical storms, and volcanic activity would have led to the formation of a "prebiotic" soup, a mixture of organic materials that would become the building blocks of primitive life.

WHOOPS! I DON'T THINK WE EVOLVE FOR A FEW HUNDRED MILLION YEARS YET!

In the 1950's, Harold C. Urey and Stanley L. Miller:...

...mixed ammonia, methane, hydrogen and water together in a large flask and to simulate the electrical storms subjected the mixture to periodic electrical discharges.

Within days amino acids began to accumulate in the apparatus!

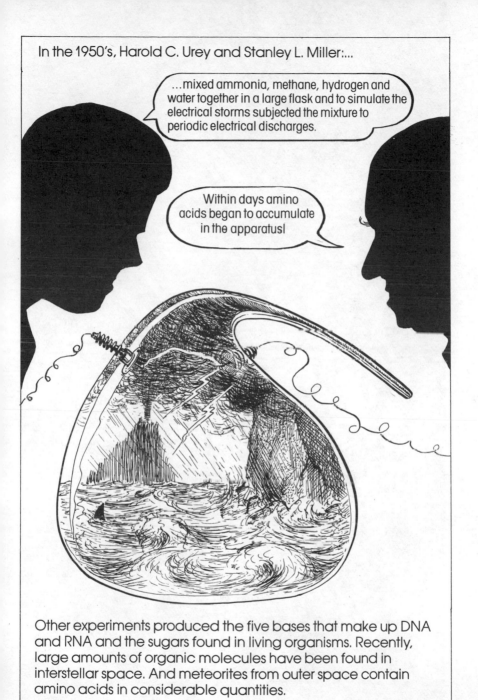

Other experiments produced the five bases that make up DNA and RNA and the sugars found in living organisms. Recently, large amounts of organic molecules have been found in interstellar space. And meteorites from outer space contain amino acids in considerable quantities.

Prebiotic soups could well be forming in many parts of the universe.

Prebiotic

CONTENTS

Yet, the essential element in the beginning of life would have been the formation of self-replicating organisms within the prebiotic soup. Scientists have shown that random chains of amino acids (proteins) and of nucleic acids (DNA or RNA) can be produced experimentally. How these chains — or polymers as they are called — develop into a system of self-replication remains unknown. Nonetheless, once a primitive self-replicating system got started it would have developed a competitive advantage.

Soup

JUST HEAT, STIR, SERVE & EVOLVE

Random aggregations of molecules and the **selection** of the more successful duplicating processes were, most probably, the driving forces in the formation of life.

The oldest known primitive organisms, found in sedimentary rocks in Australia dating from 3.6 billion years ago, probably lived off the fermentation of various organic materials. Photosynthetic processes evolved and they began producing oxygen.

For one hundred million years the oxygen produced by these organisms reacted with the iron in the oceans and precipitated the iron out as gigantic bars which form the major portion of today's iron reserves.

Only after the oceans had been 'rusted' did oxygen begin to fill the atmosphere.

The presence of oxygen forced many primitive organisms into the protective cover of oxygen-tree environments.

Those cells that could tolerate an oxygen atmosphere evolved mechanisms for using the oxygen.

A GALACTIC VIEW OF DNA

There is an interesting side to the evolutionary process that is illuminated by astronomy. The living organisms we now see all have their structure based upon the element carbon. Most biochemists believe no other basis is possible for life. But where does carbon come from? Carbon originates in the centre of stars where at temperatures of millions of degrees it is 'cooked' from simple protons and neutrons. When the stars reach the end of their lives they explode and disperse carbon into space and on to the surface of planets and meteorites. However, the time needed to make carbon and other heavier elements, like nitrogen and oxygen, by this stellar alchemy is very long: nearly a billion years. Only after this immense period of time will the building blocks of life be available in the universe, and only then can biochemistry take over.

So, life is only possible in a universe that is at least a billion years old. Remarkably, because the universe is in a state of expansion, this also means that life can only arise in a universe that is at least a billion light years in size. The vastness of the universe is inextricably bound up with the existence of life within.

John D. Barrow

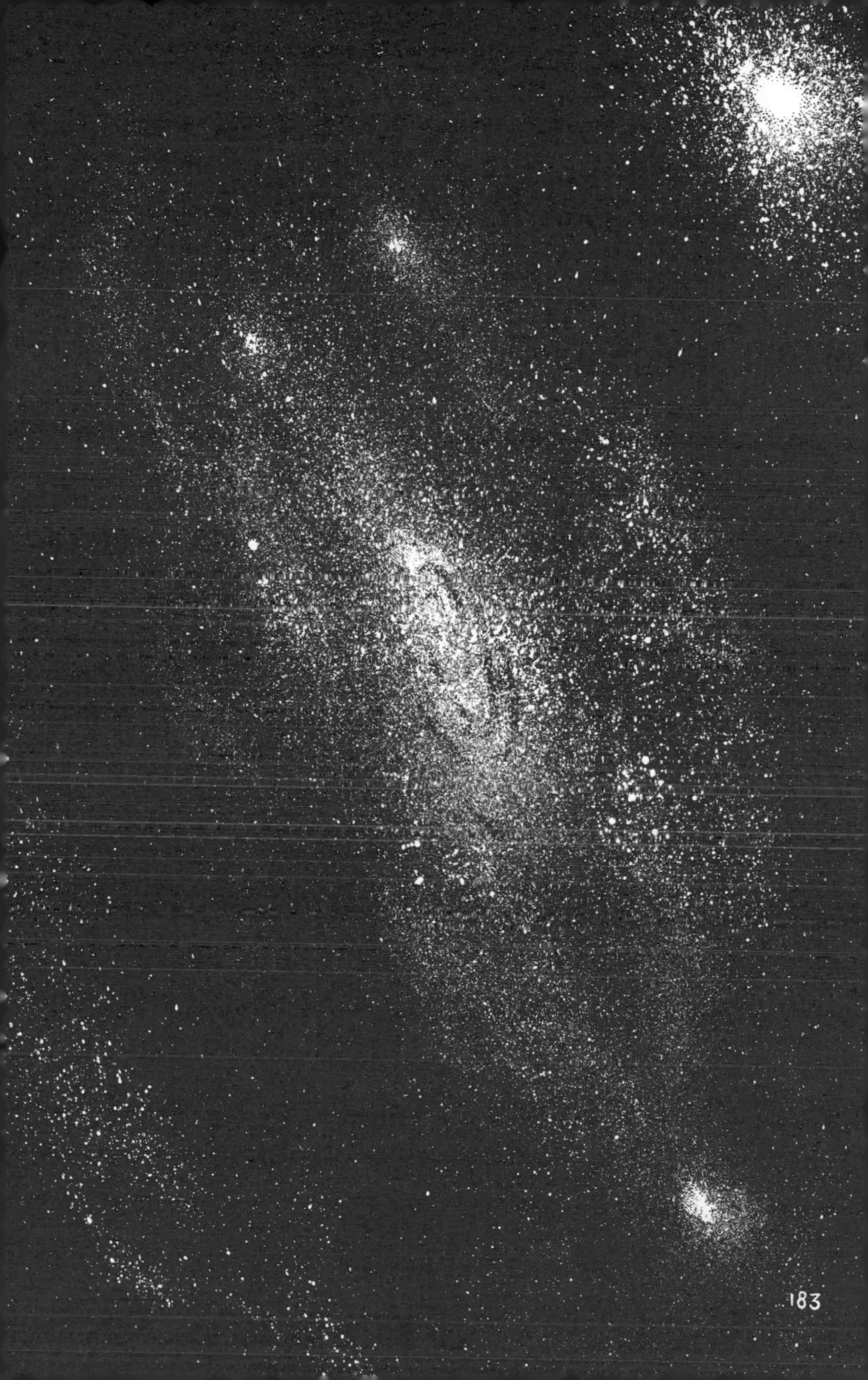

About 1.4 billion years ago the first cells with nuclei –
the eukaryotic cells (eu = true; karyote = kernel) –
appeared. Advanced sexual reproduction was
now possible with a consequent
more rapid pace to evolution.
By one billion years ago
multicellular organisms
had evolved.

MORE FUN
THAN PRIMEVAL
SOUP...

The history of the earliest life forms shows how changes in the environment created new selective pressures, giving rise to new life forms.

But what is natural selection selecting? And does our knowledge of the structure of DNA give us any insights into the possible molecular mechanisms?

Variations appear to be randomly produced. Many do not help the organism adapt to its environment. (As Gould and Lewontin have said: the "male tyrannosaurs may have used their diminutive front legs to titillate female partners, but this will not explain **why** they got so small.")

Soo... BUT THANK THE LORD THEY DID!

Natural selection does not foresee the future.

Adaptation is not like solving an engineering problem. The jet engine did not **evolve** from the combustion engine, but was built from scratch. Biological organisms must somehow incorporate what is already there into the new organism. Francois Jacob:

> Evolution proceeds like a tinkerer who, during millions of years has slowly modified his products, retouching, cutting, lengthening, using all opportunities to transform and create.

I TINKER
THEREFORE I AM

Because the globin gene – discussed earlier – was duplicated, an extra copy was available for tinkering. Mutation and natural selection could then create globin diversity.

In 1976 Richard Dawkins published his book **The Selfish Gene** creating a considerable stir throughout the scientific and even philosophical communities. Dawkins argued that selection is at the gene level.

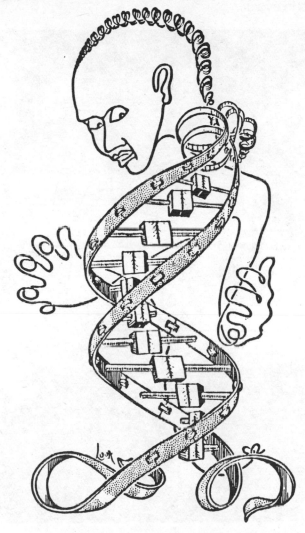

The aim of a gene, he said, is to survive from one generation to the next and it **uses** the bodies of living organisms.
Human beings are simply survival machines for DNA.

Then in 1980 Francis Crick and Leslie Orgel presented the ultimate argument for self-centred molecules: **selfish DNA**. Some DNA exists, they said, not because of any benefits it might bring to an organism, but because that DNA is what is being selected in evolution.

Not all DNA is "selfish" and the organism lets the selfish DNA exist as long as it doesn't get too much in the way, because it would take too great an effort for the organism to get rid of the 'junk' DNA.

The selfish gene (and selfish DNA) idea, with its metaphorical image of living things being manipulated by DNA, had its genesis in a paper written by the English population geneticist D.W. Hamilton in 1964. He tackled a problem that had puzzled Darwin:

The evolution of sterility in certain species of insects, (which include ants, bees and wasps).

Hamilton noted that the females of these species have pairs of each chromosome: Two sets.

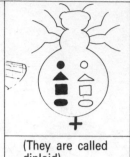

(They are called diploid).

Males however have one of each chromosome: One set.

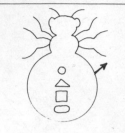

(They are called haploid).

To parent a daughter:–

The father gives his entire set– *the same for each daughter.*

The mother also gives one set, but she draws this set by taking some chromosomes from her first set and some from the second. . .

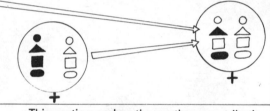

. . . This sorting makes the mothers contribution *different for each daughter.*

8 SISTERS. . . SISTERS. . . NEVER WERE THERE SUCH DEVOTED SISTERS!

When we do the sums, sisters are more closely related to each other than to their mothers, and even less closely to their offspring.

Therefore **if** genes are selfish, sisters (even sterile ones) are better off helping each other than their offspring if they want their own genes to survive. This is exactly the way these insects behave.

Of course, this doesn't prove that genes **are** selfish.

Curious, isn't it, though, that the ants and bees behave just as you would expect them to if they were being controlled by selfish genes?

It would be foolish to draw conclusions about human behaviour based on analogies with insect behaviour (which is largely programmed by their genetic make-up). Human behaviour is determined by genetic factors against a powerful background of cultural and moral beliefs and relationships. An ant could never avenge an ancestor's death, believe in God ... or discover DNA!

When the structure of DNA was first elucidated in 1953 it was believed that random mutations in the DNA structure and sexual recombination would account for evolution. Genes often exist in duplicate copies in an organism and the process of duplication allows for the creation of mutant structures — that may or may not help the organism adapt — without sacrificing the original gene.

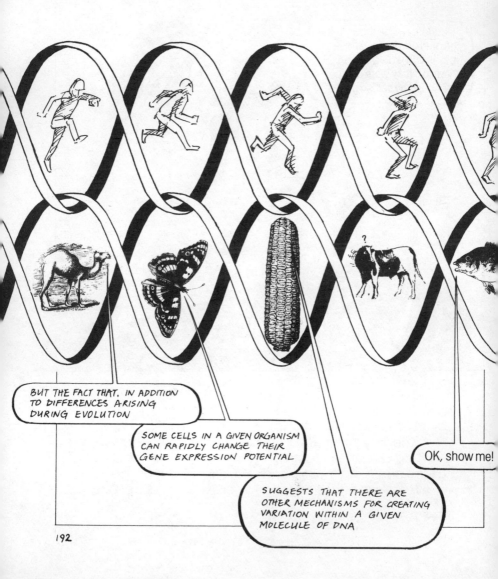

BUT THE FACT THAT, IN ADDITION TO DIFFERENCES ARISING DURING EVOLUTION

SOME CELLS IN A GIVEN ORGANISM CAN RAPIDLY CHANGE THEIR GENE EXPRESSION POTENTIAL

OK, show me!

SUGGESTS THAT THERE ARE OTHER MECHANISMS FOR CREATING VARIATION WITHIN A GIVEN MOLECULE OF DNA

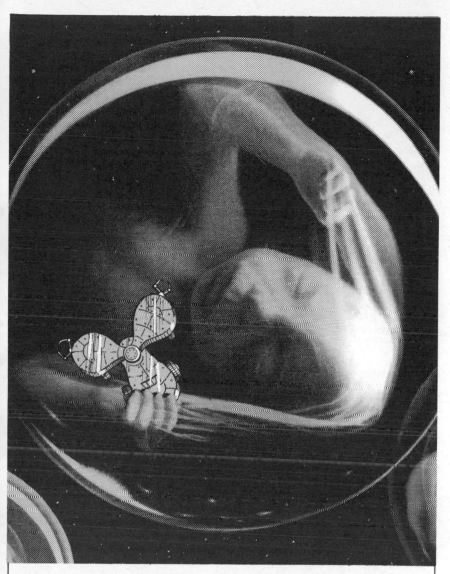

It was soon learned that there are mechanisms for reassorting DNA within a particular plant or animal. One of the most remarkable of these is responsible for the production of millions (at least) of **antibodies** from a few hundred antibody genes, permitting man to survive when infected by new kinds of organisms. The stock of possible antibodies is determined during embryonic development.

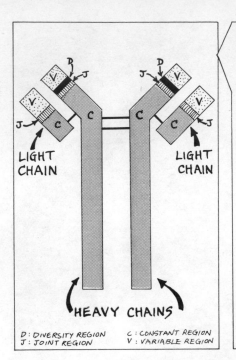

D : DIVERSITY REGION   C : CONSTANT REGION
J : JOINT REGION   V : VARIABLE REGION

Antibodies consist of two types of protein chain: light and heavy.

Each light chain has variable and constant regions. The variable consists of two segments: V and J. The constant is called C. Heavy chains are similar except their variable region is in three segments: V, D and J; as well as a constant C region.

The remarkable property of the immune system is its ability to create an immense number of antibody types. The variety results from the differences in the variable regions in the light and heavy chains.

How does this arise?

In the embryonic DNA we find the genes for these various segments. But there are many of each segment type and they are widely separated in the DNA, that is, **not yet assembled as a functional gene**. To make the gene for a light chain or a heavy chain, the distant DNA segments must be brought together and joined.

There are many possibilities for this joining:

| ABOUT 200 V GENES | ABOUT 12 D GENES | ABOUT 4 J GENES | ONE C GENE FOR FIRST JOINING STEP |

Here's a heavy chain family.

In antibody producing cells, an active **heavy** chain gene is made by joining one D gene to one J gene to make a DJ (12 x 3 = 36 possibilities), and then the DJ to a V gene to make a VDJ gene (36 x 200 + 7200 possibilities).

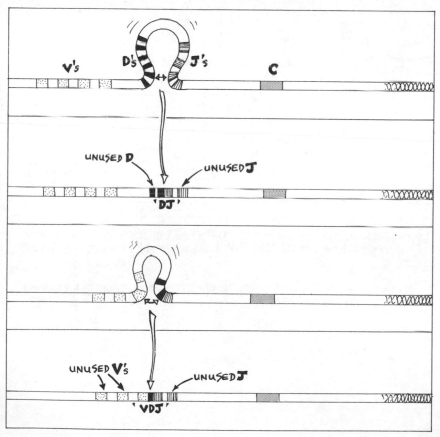

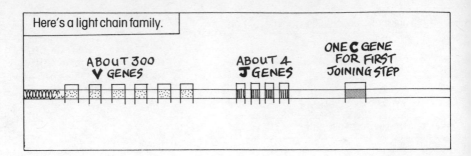

Here's a light chain family.

ABOUT 300 V GENES

ABOUT 4 J GENES

ONE C GENE FOR FIRST JOINING STEP

For lights chains a similar DNA rearrangement brings V to J. The possibilities for different light chains are about 300 x 4 = 1200

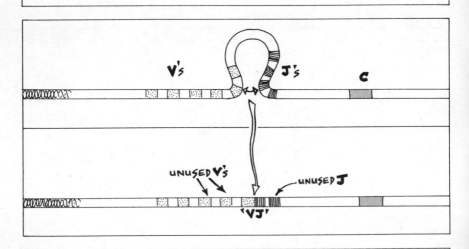

V's   J's   C

UNUSED V's   UNUSED J

'VJ'

Even after such DNA rearrangment the **VDJ** for heavy chains is separated from the heavy chain **C** gene; also the **VJ** for light chains is separated from the light chain **C** gene.

For both heavy and light, **C** is brought into place **after** transcription, by the **splicing** step which forms the mature mRNA.

For light chains:

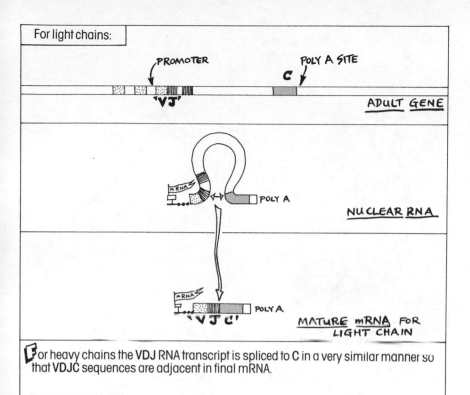

PROMOTER

POLY A SITE

C

'VJ'

ADULT GENE

mRNA

POLY A

NUCLEAR RNA

mRNA

POLY A

'V J C'

MATURE mRNA FOR LIGHT CHAIN

For heavy chains the **VDJ** RNA transcript is spliced to C in a very similar manner so that VDJC sequences are adjacent in final mRNA.

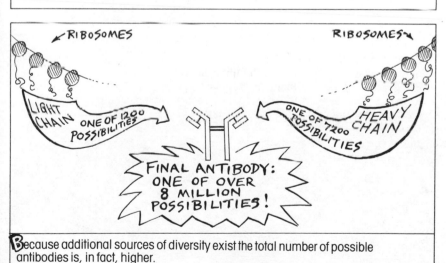

←RIBOSOMES

RIBOSOMES→

LIGHT CHAIN ONE OF 1200 POSSIBILITIES

ONE OF 7200 POSSIBILITIES HEAVY CHAIN

FINAL ANTIBODY: ONE OF OVER 8 MILLION POSSIBILITIES!

Because additional sources of diversity exist the total number of possible antibodies is, in fact, higher.

Mature mRNA's for light and heavy chains are thus translated.

After translation in the ribosome, the light & heavy protein chains form an antibody . . .

. . . which is sent to the surface of the cell which produced it.

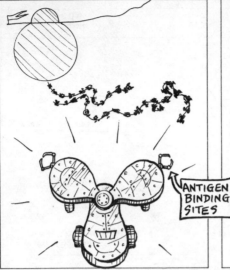

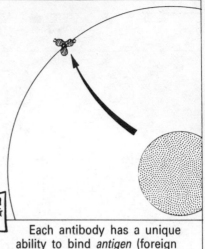

Each antibody has a unique ability to bind *antigen* (foreign substances which enter the bloodstream).

Perched on the cell surface the antibody surveys for any antigens it can bind.

Each blood cell has recombined its antibody genes in a slightly different way, making different antibodies which bind different antigens. Remarkably, binding of antigen triggers the producing cell to proliferate to tens of thousands of identical cells, each spewing antibody out into the bloodstream!

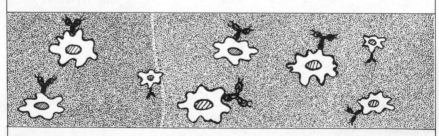

In this way, reassorting antibody genes creates diversity, and antigen binding selects the necessary response.

Mechanisms similar to those responsible for antibody production may be important in cell differentiation and the production of diversity in individual organisms. Transposable elements (jumping genes) have been found that can insert themselves into a variety of sites in DNA, causing mutations, inversions and the turning on or off of genes.

Even the so-called "junk" – the DNA in the introns that does not appear to have any clear function – may eventually be found to play an important role in regulation and the production of variation.

NEWLY-PLACED
GENE

In the 1940s, Barbara McClintock discovered "jumping genes". Jumping genes are capable of moving from place to place within the chromosome, and inserting themselves between or within other genes. In some cases genes which neighbor the insertion site may be turned on or off when a new gene jumps into their vicinity. When the jumping gene is inserted within another gene, it can alter the coded information of that gene. Jumping genes were first discovered by McClintock in maize, and it is now known that certain genes in animal cells, such as the immunoglobulin genes in mammals, also "jump" through highly controlled DNA rearrangements. Some DNA sequences jump by making copies of themselves which can insert at other sites in the chromosome. These DNAs "proliferate" within the chromosome through the jumping mechanism, and may be found repeated hundreds of thousands of times within the genome. Because this proliferation may not be of direct use to the organism itself, these DNAs have been called "selfish DNA".

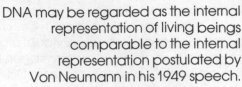

DNA may be regarded as the internal representation of living beings comparable to the internal representation postulated by Von Neumann in his 1949 speech.

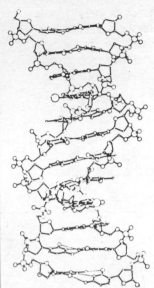

But, it is a complex and unusual representation indeed. Although a man may design a machine in a single day, DNA has achieved its form and complexity through eons of evolution, reaching back to the earliest life forms. In this sense DNA is an inheritance which joins us to our most remote ancestors in the past.

Quite a remarkable molecule indeed.

ends

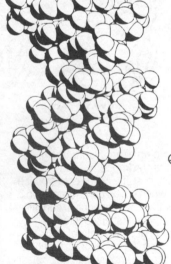

SKELETAL & SPACE-FILLING MODELS OF THE 'B' FORM OF DNA

Post-Script

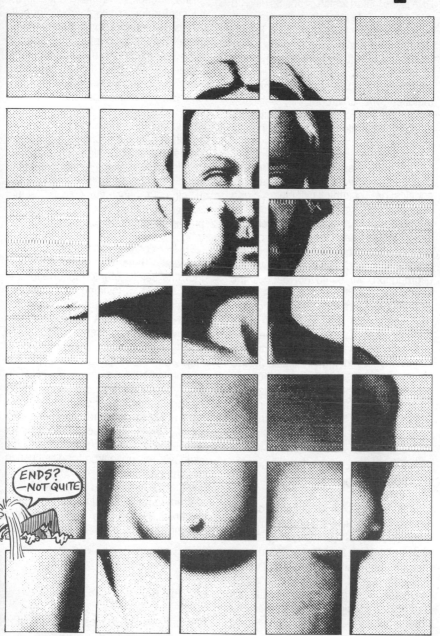

DNA — Our Most Valuable Heritage

DNA is the thread which connects us with our most remote ancestors. If there were any interruption in the chain of inheritance of genetic information, the "evolutionary value" of previous millenia would be lost. The coded genetic information in DNA is the outcome of mutation and environmental selection which together create the evolutionary process. This information could not be replaced.

If we knew every detail of the structure of a cell, apart from DNA, the chemical constitution of the cytoplasm and nucleoplasm, the lipid content of the membranes, the amino acid sequence of every protein and the folding of the peptide chain, the energy levels of the metabolites and their pathways of degradation, we could not deduce the structure of the DNA. Yet, the DNA of the diploid chromosomes in the undifferentiated egg is sufficient to determine the minute details of the developed organism. By saying that a hen is only an egg's way of making another egg, we emphasize the requirement that as generations pass, the DNA must be transmitted. Germ cell must give rise to germ cell, for a species to continue to exist.

A Universal Code

How is our genetic information related to the information of other species? We can clone the genes of many organisms and attempt to rank them on an evolutionary scale. The first fact which is clear is that the genetic code is universal. The genes of virtually all organisms "speak" to the ribosomes and tRNAs of the protein-synthesizing machinery in the same genetic language.

One exception is the code of the mitochrondrion, a cytoplasmic organelle with its own chromosome. Mitochondria are believed to be captured organisms, originally residing symbiotically in the cytoplasm of their host, and now essential components of eukaryotic cells. But even the mitochondrial code is nearly identical with the conventional genetic code. One other possible exception is the *prion*, the protein component of the infectious agent of scrapie, which causes a fatal disease in sheep. The scrapie agent has no evident nucleic acid, hence the speculation that its genetic information is caried in "protein genes" called prions.

These cases aside, the code is universal. There are many variations upon the details of information storage, with viruses providing the greatest deviation. The genome of poliovirus is a single-stranded RNA which serves directly as a messenger RNA. And influenza virus uses helical rods of double-stranded RNA to store its genetic information, with a separate rod for each gene. Provocatively, scientists using cloning technology have made DNA equivalents to poliovirus and influenza genes. When introduced into animal cells, these homologues direct the synthesis of poliovirus and influenza proteins, without any concern that the "stuff" of the original genes was RNA.

What remains constant through these examples is the code itself, which is common to man, bacteria, flu and polio.

Our Ancestry

This universality suggests that all living

things had a common ancestor which bequeathed the code to all of today's living organisms. A gene from a fruit fly need never function in a human cell (although scientists have recently made them do so!) However, flies and men, and their respective ancestors were endowed with a common genetic code, and too much was at stake (i.e., the expression of their genes) for the codes to change during evolution.

If we dare to speculate about a common ancestor for all living entities, then surely we can rank existing beings, bacteria, fruit flies, man, etc., on some evolutionary scale. Does man, and the animal kingdom represent the triumph of evolution, when compared, say, with the lowly E. coli bacterium?

Bacteria: Relic or High Tech

Conventional wisdom placed the simple bacterium at the bottom of the evolutionary scale. Man has placed himself at the top. The physicist Eddington postulated that "entropy is time's arrow". Using this postulate, we can distinguish the progression of time. For example, if we had a filmstrip showing a building disintegrating, we would know that we were playing the filmstrip forwards by observing the building falling apart as time progressed, not reassembling from bits of plaster and timber to form an intact edifice. Is there a similar rule we can apply to determine the direction and the progress of evolution?

Although bacteria seem lower (hence more ancient) than the multicellular eukaryotes, they have some remarkable properties. E. coli can reproduce in twenty minutes, about 350,000 times faster than the generation time of man! Bacteria lack the intronic sequences found in eukaryotic genes. They lack the splicing mechanism which removes the intronic sequences during processing mRNA, and they lack the nucleus as well. They possess a single chromosome which is in the cytoplasm. In fact, they lack most features of eukaryotes which make day-to-day life, and the generation of progeny, time consuming. Thus a view which opposes conventional wisdom is that bacteria are highly evolved creatures, "streamlined" for a rapid life style. If bacteria evolved from eukaryotes the streamlining would required that introns be removed from every gene (it would be too much work to evolve the genes from scratch) so that the emerging new prokaryotic life form could dispense with splicing and the nucleus. Indeed, we now know that eukaryotic genes *can* loose their introns by a mechanism in which the mRNA (an "intron-less" form of the gene sequence) is copied back into DNA. Such intronless genes appear to be fairly common in animal cells, and reverse transcriptase (which we encountered in cDNA cloning) or its equivalent could do the job for converting mRNA information (without introns) back into "streamlined" DNA. Evolution of prokaryotes from eukaryotes is not far fetched!

How Organisms are Related: The Clue in DNA

If we doubt the ranking of bacterial prokaryotes and eukaryotes in the evolutionary scale, can we ever hope to establish the direction of evolution and thereby glimpse our own origins? Perhaps when the structures of the genes of many organisms have been determined by DNA sequencing (only a million nucleotides are now known) we will be able to perceive a clear structural progression from one species type to another, which will rank these organisms chronologically in the order of their appearance on the evolutionary stage. However, recent evolution of a species, and even diversity of biological function (such as playing the cello or reciting Chaucer) are not guarantees of evolutionary fitness. Perhaps after a man-made nuclear holocaust, only cockroaches would survive. A humiliating test of our evolutionary rank!

The Individual and the Species: The Consequences of Mutation

From generation to generation, the adult members of the species die, the differentiated somatic cells are lost, and the informational inheritance in DNA is passed through the germ cells to the next generation.

The species and the individual have somewhat conflicting interests in the transmission of DNA's information, especially in the fidelity with which it is transmitted. The stability of genetic information is in the best interest of the individual. For the individual to survive, he must be well constructed and in good running order. Assuming that the parents of the individual were themselves free from disabling genetic defects, it is in the offspring's interest that he inherit DNA copies of the parents' genomes that are of the highest fidelity. Changes in the nucleotide sequence of the DNA, called mutations, can occur through chemical or x-ray induced damage to the DNA chains. Mutations also arise through base pairing errors during replication, or through chromosomal rearrangement, that is, transposition of DNA segments

from one chromosomal location to another.

The Individual: "No Errors Please"

To avoid mutational errors, cells employ proofreading functions, built into the DNA polymerase, that scan the newly synthesized DNA to edit base pair errors and correct them. Specific enzymes remove nucleotides damaged by radiation. Thymine dimers, a common example in which neighboring thymine bases in a chain are linked together, are excised and replaced. Despite these precautions, errors occur. Mutations can affect any gene or DNA segment, and alter its expression in unexpected ways. Many changes will have virtually no impact on the individual's ability to survive. A nucleotide change in a DNA region between genes, that does not code for protein, might have little or no effect on the organism's function. However, nucleotide changes in functional genes will, in general, be deleterious. If a computer manufacturer arbitrarily changed the specifications for the wiring of a circuit board (akin to a random mutation) substituting resistor for capacitor, or transistor for micro-chip, the product most likely would not function. For complex machines, like computers or man, *random* changes will probably be for the worse.

Species: 1) Tinkering Permits Variation

The mutability of genetic information is essential to the survival of the species. To continue our analogy, a *very, very* small number of random changes in the design may actually improve the computer. Most computer manufacturers enforce strict quality controls to ensure error-free construction, and employ electrical engineers and market analysts to make carefully planned changes which improve product design. In contrast, nature does not avail itself of analysts and engineers to evolve DNA blueprints. However, the need to evolve remains great. Thus, the species must tolerate tens of thousands of genetic errors imposed upon offspring, causing marginally lower survival value, to obtain, by chance, the one slight genetic improvement that increases survival value.

2) Introns and Evolution

The existence of exons and introns, and the splicing mechanism, provides additional means for evolution. DNA exon segments transposed into introns of genes can add protein coding sequences to mRNA by means of the splicing mechanism. Hence new peptides may be inserted into protein. Should such a new exon encode a functional domain of a protein, this process of "exon shuffling" could provide old proteins with new activities, in a small number of evolutionary steps. Splicing also permits "tinkering" with protein structure at the RNA processing stage of gene expression by varying the coding sequence retained in the mRNA product.

Mutations can arise by many mechanisms. The species as a whole relies on the mutability of its genetic information to permit evolution. However, the individual relies on the invariant, fully faithful transmission of the same information in germ line cells, to ensure freedom from genetic disability.

3) Cancer and the Genetics of Somatic Cells

Germ line cells provide the DNA for future generations. However, for the individual, genetic transmission also occurs through *somatic* cells. Although somatic mutations have no direct genetic consequences for future generations, our new understanding of cancer's origins show that they can have profound consequences for the individual. Cancer is a somatic genetic disease, an example of natural selection amongst somatic cells, working against man.

Once we reach adulthood, the number of cells in our body remains relatively constant. Some cells, such as neurons, which are "terminally differentiated" and non-dividing may continue to function for the lifetime of the individual, and will not be replaced. Other cells, such as red blood cells, have a finite lifespan. They are routinely inspected in the spleen for inperfections and removed from the blood stream if found faulty The "hemopoeitic progenitor cells" in the bone marrow retain the ability to divide, and (unlike neurons) produce daughter cells which differentiate to give new red blood cells, thus replenishing the population.

Control of cell division is a crucial feature of the society of cell types which comprise the mature organism. Cells must proliferate only in response to carefully pre-programmed signals. When cell division proceeds unchecked, outside the normal requirements for cell replenishment, a tumor results. If the tumor cell mass disseminates or impairs the function of vital organs, the resulting cancer is life threatening. We now know that many cancers result from somatic mutations which provide cancer cells with an (undesirable) ability to proliferate. Mutations in somatic cells which allow proliferation will, through a cruel form of natural selection, allow the mutant cell type to thrive at the expense of its neighbors and ultimately at the individual's expense too.

Mutations That Cause Cancer

Scientists have used growth selection to identify the genetic changes which convert normal cells to cancer cells. DNA extracted from tumors was introduced into cultured benign cells. Those which took up and expressed the gene for cancerous growth proliferated in the tissue culture dishes of the experimentalists, and are said to be "transformed". This unchecked growth of transformed cells mimics tumor growth in the organism.

The newly acquired gene, called an *oncogene*, which originated in the tumor, and which transformed the cultured cells, was isolated by cloning and the protein it encoded was identified. Normal tissues were found to have counterparts to the oncogene, and these were called proto-oncogenes. Proto-oncogenes were also isolated by cloning. Several normal (proto-oncogene) and cancerous (oncogene) gene pairs were identified in this manner. For the best studied example, called the *ras* gene, the DNA nucleotide sequences of proto-oncogene and onco- gene were determined and compared. The *ras* oncogene from a colon carcinoma had a single nucleotide change in comparison with its normal proto-oncogene counterpart. A

"G" residue in the codon for the twelfth amino acid of the *ras* gene protein had changed to a "T" residue. While the proto-oncogene specified a protein whose twelfth amino acid was a glycine, the oncogene made a virtually identical protein, but with valine at position twelve.

Why was this nucleotide change important? Both normal and transformed cells make similar *quantities* of the *ras* gene protein. Thus a change in the level of expression is not implicated in cell transformation. Instead, the oncogenic potential of the mutant gene is likely to reside in its specification of an altered protein, with a mutant structure. Investigation of other tumors also revealed cases in which the oncogene was a *ras* gene. In each case, the change in *ras* gene DNA structure, which converted a normal *ras* proto-oncogene to an oncogene, resided in codon twelve. Thus, in other examples, glycine was replaced by aspartate, by serine or by lysine.

We don't fully understand the function of the *ras* gene protein in healthy cells, or why its mutation causes uncontrolled proliferation. Apparently the *ras* protein plays a crucial role in cell growth control. When glycine, a small amino acid, is replaced by any of several bulky ones, the *ras* protein function may be impaired. Most likely the mutant protein chain cannot fold to assume a fully functional structure. Like a minute engineering error which causes an automobile brake to fail, the small change caused by the *ras* oncogene mutation can lead to disaster!

The recent discoveries of oncogenes follow directly from the technological advances in genetic engineering and DNA manipulation. Through implementation of gene transfer, tissue culture selection, DNA cloning and sequencing, we can lay bare the structure of any gene or its mutants, and predict the sequence of its protein product. Ironically, we can generate this new information more rapidly then we can assimilate and interpret it. Thus, we know that the mutant *ras* oncogene is a "cancer criminal". We are still unsure of the exact nature of the crime against cell growth regulation that the mutant *ras* has committed.

Are We More Than What DNA Has Made Us?

Since it is now possible, in principle, to decipher the genetic structure of any organism, we can ask ourselves to what extent is the organism determined directly by the DNA? For example, is DNA complexity directly related to that of the organism? While earlier studies on bacterial genes gave the impression that most bacterial DNA codes for protein, scientists were quite surprised and even puzzled by the discovery that for animal cells, only ten percent of the DNA is protein coding. In fact, the function of most of the DNA in animal cells remains unknown, and some scientists have suggested that it doesn't have any functional role at all. Whatever the case, we can determine the total quantity of DNA in cells from different organisms and relate this to morphological complexity. We find that a frog cell has 3.5 picograms of DNA, a human cell 3.4 picograms, and that of a lilly flower 32.8 picograms. From these figures, we see that DNA content is not a simple index of the structural complexity of an organism. Furthermore, a mouse brain has some six million cells, whereas a human brain has tens of billions of cells. There is no hint of this in the DNA content of these cells.

We have already asserted that the DNA of an organism, transmitted through the germ cells, determines the final morphology of the adult. If we knew all of the structural details of the genes of an organism, including the complete nucleotide sequence of the DNA, would we be able to predict this morphology? How in fact does DNA act to determine the structure of the organism?

We know that DNA determines the linear sequence of amino acids in the protein translation product. Yet, the essential characteristics of proteins depend not just upon their amino acid sequences, but also upon the detailed three dimensional structure of the folded protein chains. Subtle physical and chemical interactions between different amino acids of the chain, and between the chain and it chemical environment, determine the precise nature of the folding. Although we can deduce the linear primary sequence of amino acids from the DNA, to calculate the way the protein folds is at the limit of our current capabilities. Second, if we knew nothing about the gene save its structure, we might have great difficulty deducing the function of the protein product, for example, its catalytic activity, were it an enzyme.

If we consider that any cell has tens of thousands of different proteins, and that the body is composed of tens of billions of different cells, the computational problem becomes immense. Finally, to deduce the various metabolic reactions and activities which constitute the normal routines of the cell would be immeasurably beyond our capabilities. Thus, despite the great power of the "new biology" and its techniques, it alone is insufficient to unravel the molecular mechanisms of development.

Fortunately, more traditional genetics provides us with some insights. A simple way to study development is to analyse organisms with mutations in specific genes that control differentiation, and to note the consequences in the adult organism. We will find that what is altered is more complex than we would have predicted from simple bacterial models. Nature has already done such experiments for us, and one example is the condition called *albinism*.

Genetic Case Studies: 1) Albinism

Among the striking features of this condition are light skin and red eyes in humans, or the distinctive pattern of light and dark patches in the Siamese cat. The site of the mutation is in a gene believed to determine the structure of an enzyme, *tyrosinase*. Tyrosinase plays a crucial role in the synthesis of the pigment melanin. The light color of the skin (or fur, in cats) is a direct consequence of the absense of melanin. But that is only part of the story. Melanin also is found in the epithelial layer of the eyes. Albinos lack the pigment and have red eyes. For reasons that are poorly understood, the failure of melanin to appear in the epithelial layer behind the retina, causes the optic nerve (which normally partly crosses as it projects to the brain) to project in an abnormal manner. (This is seen in Siamese cats as well as humans.) Albinos have either very restricted visual fields and/or only monocular vision. What is apparently a simple mutation affecting skin pigmentation causes a cascading effect which includes severe neurological problems. A single gene has consequences for many aspects of an organism's development. In keeping with this complexity, the gene that codes for the enzyme tyrosinase, cannot be

said to code directly for vision, or for that matter, for the circuitry of the brain.

2) Genes Versus Environment in Sexuality

Animal behaviour also depends on both genetic and environmental factors. For example, the development of animal and human sexuality is not as clearly 'genetically determined' as one might believe. The hormonal environment of the developing foetus can have profound effects on later sexual development and activity.

The characteristic position of the female rat during sexual intercourse is known as *lordosis*. The female arches her back and sweeps her tail aside in order to receive the male sex organ. Males, on the other hand, exhibit mounting behaviour. Male rats that are castrated at birth will exhibit lordosis as adults rather than mounting behaviour. Therefore, a rat that is genetically male will behave as a female. This can be prevented by giving the rat injections of male hormones during the period immediately following its birth. If the hormones are given later in life, they will fail to suppress the lordotic posture in intercourse. In the female, lordosis is eliminated when the ovaries are removed.

These crude experiments only suggest the subtle variations in sexual behaviour that result from differing hormonal environments of the foetus. Female and male brains have been found to be anatomically and neurochemically different and these differences develop during the *in utero* period. Female brains are modified when male hormone is present in the uterus because of a male litter mate.

Genetic Programs for Behaviour

While environmental factors, be they *in utero* or post-natal, certainly profoundly affect the manner in which genes are expressed, rather complex, apparently genetically determined patterns of behaviour have been known for some time.

The Dutch ethologist, Nikolas Tinbergen, now living in England, showed many years ago that some animals engaged in certain relatively fixed patterns of behaviour. In one famous example, the Stickleback fish, Tinbergen demonstrated that the mating ritual followed a fixed sequential set of acts, each aspect of which was initiated by a definite physical signal. The male Stickleback, for example, acquired a red spot on its belly during the mating season. The female will follow a male with this sign to its nest. Any more or less fish-shaped object with a red spot could lead the female on. If the object be turned upside-down, so that the spot now appeared on the top, the female fails to follow. The entire mating procedure in the Stickleback required clear signs (such as the red spot on the belly of the male) presented in the proper order. Whether these signs are produced by the mating fish or a mechanical object, does not matter. The appearance of the sign produces the appropriate behaviour. The implication of Tinbergen's work was that this behaviour is not learned but rather is genetically controlled.

Recent studies on Aplysia, a shell-less marine snail which grows to the size of the human brain, have in fact isolated some of the genes that control mating behaviour in that animal. Egg laying behaviour consists of a specific series of actions, a rigid behavioural pattern which is phenotypic of the species: copulation, ejecting an eggstream from a duct, increasing heart beat, waving the head, catching the eggstream in its mouth, winding the stream into a solenoid, and placing it on a rock. Peptide hormones, several amino acids long, produced in the "bag cells" of Aplysia will induce this behaviour if injected into a virgin animal. The laboratories of Richard Axel and Eric Kandel cloned the gene for the egg laying hormone. From its DNA sequence they deduced that the peptide hormones responsible for this behaviour are produced by cleaving a larger protein which is synthesized in the bag cell. Their work suggests that different peptide hormones cleaved from the larger peptide activate different steps in the egg laying ceremony. Thus a gene directly determines the onset of these actions, with different segments of the gene apparently responsible for different components of the activity.

Behaviour Beyond the Reach of Genes

These examples of genetic determinism far from prove that all behaviour is genetically controlled. Genes do not operate in a vacuum. The Stickleback's rituals are initiated by different environmental cues. If they fail to appear, so will the ritualized acts. Of course, in principle, we could alter the nature of the performance by altering one or two genes, as might be possible with Aplysia. But most behaviour is not the consequence of a specific gene. Our linguistic ability certainly has a genetic basis in the organization of the brain, but the language that we speak is not determined by our genes. Children of English-speaking parents are not born with "English language" genes anymore than children of Japanese parents are born with "Japanese language" genes. Rather we are born with a capacity to learn *any* language to which we might be exposed. There are no genes for a specific langauge, and likewise there are no genes for specific thoughts either. Language and thought — thought dependent on a genetic *capacity* to learn — are also the consequences of environmental factors which cannot be programmed genetically.

Bioengineering Cures for Genetic Disease

There are specific mental disorders (such as schizophrenia) that may be the consequence of physiological dysfunction. Such diseases are today controlled through medication, with varying results. Part of the problem is that we are not as yet sure just what the physiological malfunction is, nor whether it has a genetic basis. But in the cases that diseases have a clear genetic cause, would engineering be of any use in curing the disease? Sickle cell anemia is an example of a disease in which nucleotide changes in the beta globin gene yield a mutant globin protein, with altered oxygen binding properties. Does the new technology provide us with any procedures to correct malfunctioning genes?

We are still at the early experimental stages of clinical applications, but a recent experiment indicates some of the future possibilities. In late 1982, a team of scientists reported that they had produced mice that were almost twice

normal size.

Growth hormone is a protein that has profound effects on the development of cells, and among the more deleterious effects of abnormally high levels is the production of giantism in humans. The scientists therefore decided to see if they could introduce the gene for growth hormone into a mouse embryo, activate the gene, and produce an abnormally large mouse. Since the control regions of the growth hormone might have prevented its activation in most mouse tissues, the gene was attached to the transcription control region of the mouse metallothionein gene — a gene that is, turned on by the presence of toxic heavy metals. The advantage of this promoter control region is that it is active in most tissues of the mouse, though the level of expression varies from tissue to tissue. The assumption, then, was that the promoter for the metallothionein gene promoter would be active in most of the cells of the body of the mouse and that it would in turn activate the growth hormone gene to which it was attached.

The scientists injected the metallothionein promoter linked to the growth hormone gene into the pronucleus of a fertilized mouse egg and then inserted the egg into the reproductive tract of a foster mother. The foreign DNA fragment was apparently integrated into the embryo genome, and because its promoter was from the metallothionein gene, the growth hormone gene was effectively expressed, resulting in a 'giant' mouse.

The success of these experiments suggests the possibility of new therapeutic procedures for correcting genetic defects as well as some perhaps more questionable ones of creating giant cows, fowl, etc. But it also points to the limitations of our present technology.

Growth hormone was expressed throughout the body of the giant mice. Many genetic defects are specific to particular cell types and at present there is not way of turning genes on within a given category of cells. A way might yet be found, but then present techniques require that the 'corrected' gene be introduced into the embryo. This would appear to limit its clinical usefulness, since we are unlikely to know enough about the genome of an unborn child to want to manipulate its genes while it is still an embryo at the earliest stages of development. The technique may eventually prove useful in the treatment of general somatic disorders of genetic origin, though there are many problems yet to be solved.

There is no way of knowing how future discoveries will modify our present understanding of the nature of DNA. Science is a bag of surprises. And it is the surprises that maintain our endless curiosity.

However, DNA is only part of the story. It contains the code for the linear sequence of amino acids that make up protein structure. But the clues it provides about the morphological and chemical characteristics of the organism are obscure. The fossil bones of dinosaurs tell us of the morphology of that long extinct species. Should one day a preserved piece of DNA be found among those fossilized remains, given our present knowledge and technology, we could hardly recreate the beasts. Indeed, if nature had left us the DNA and not the fossils, we would not have been able to imagine a dinosaur and we would have had no clue about its size. At best, we might have been able to clone a dinosaur globin gene and compare it to other modern globin genes.

Through DNA we have revolutionized our understanding of life. We have opened the door to new technologies that may be of great benefit medically, and even industrially. However, DNA on its own is but the plan without a context. Complete organisms are far more complex than DNA. The discovery of DNA has excited our curiosity and stimulated our desire to know more. An incredible molecule, for sure, it continues to raise as many questions as it has answered.

212

GLOSSARY

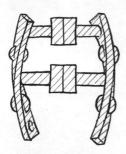 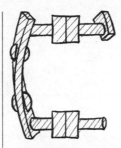

AMINO ACIDS: The fundamental building blocks of proteins. There are twenty different amino acid types (for example glycine, alanine, lysine) which are linked together during protein synthesis on the ribosome according to the coded genetic information in messenger RNA. The link which joins one amino acid with the next in the protein chain is called a peptide bond.

ANTICODON: Triplet of bases in transfer RNA which pairs with a codon in mRNA. (see CODON).

CHROMATIN: The complex of DNA with protein which resides in the living cell. In eukaryotes, the fundamental structural unit of chromatin is the nucleosome. (See NUCLEOSOME).

CHROMOSOME: A large chromatin structure thought to consist of a single highly folded DNA chain, complexed with basic proteins. In eukaryotes, chromosomes condense during mitosis into distinct X shaped structures and segregate, as the cell divides, independently amongst the daughter cells.

CODON: Triplet of bases in mRNA that code for an amino acid.

COMPLEMENTARITY: The relationship of the DNA sequence of one strand of a double helix to the sequence of the other strand. When G in one strand faces C in the other, and A faces T (or U in RNA), as dictated by the base-pairing rules, the two chains are said to have complementary sequences.

cDNA: Abbreviation of complementary DNA. It is a DNA copy of an RNA (often mRNA) synthesized by reverse transcriptase.

CONSTITUITIVE MUTANT: A class of mutants of a regulated gene that synthesizes the gene product whether or not the inducer is present. (See REPRESSOR).

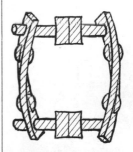

DNA HELIX, A, B AND Z FORMS: DNA can assume different double helical structures depending upon the solvent conditions and the nucleotide sequence. In the A form, which is favored at low humidity, the helix is right-handed, but the plane of the bases is inclined with respect to the helix axis. At higher humidity, and most likely in the living cell as well, the prevailing structure is the B form, also a right-handed helix, but with the plane of the bases nearly perpendicular to the helix axis. When the DNA sequence alternates between purines and pyrimidines (such as - G C G C G C G -) a left-handed helix called the Z form is stable.

DIPLOID: As applied to a cell, possessing two complete sets of chromosomes. (See HAPLOID).

DOMAIN: A structural segment of a protein molecule created when the peptide chain folds.

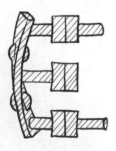

ENZYME: A protein molecule which catalyzes biochemical reactions. Examples are beta-galactosidase, which catalyzes the hydrolysis (cleavage with the addition of water) of specific bonds in sugars called beta- galactosides.

and RNA polymerase, which catalyzes the linkage of ribonucleotides to one another to make an RNA chain. Enzymes differ from man-made catalysts in that they exhibit exquisite specificity in the reactions that they catalyze, and that they function under physiological conditions.

EPIGENESIS: The doctrine that the development of the body is determined by the interaction of the genes with the environment. It holds that DNA is the plan for the organism, but that how the organism develops depends upon environmental factors.

ESCHERICHIA COLI: A bacterium found in the gut, abbreviated E. coli.

EUKARYOTES: Organisms whose cells have nuclei. These include plants, animals, protozoa, and fungi. (See also PROKARYOTES).

EXON: A continuous coding segment of a eukaryotic gene. Many eukaryotic genes are "split" and have exons interspersed with nonsense DNA called introns. Thus, a part of the gene which encodes protein. (See also INTRON).

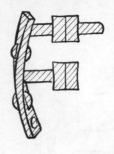

F-FACTOR: A piece of DNA that confers "maleness" upon a bacterium.

FERTILIZATION: The fusion of the sperm and the egg.

FRAME-SHIFT: The deletion or insertion of one or more bases in the coding region of a gene which causes incorrect triplets of bases to be read as codons.

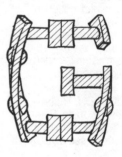

GENE: A sequence of DNA which codes for a functional product. Most genes encode proteins, but RNAs such as tRNA are also encoded by genes. Genes are the basic units of heredity.

GENOME: The total genetic information of a

cell or virus, as represented by its DNA.

GENETIC CODE: The code used by living organisms to store genetic information, in which triplets of bases in DNA (or mRNA) represent amino acids in proteins.

GENOTYPE: The genetic make up of an organism, as found within the genome. (See PHENOTYPE).

GERM CELLS: Cells that give rise to sperm and egg and thus transmit genetic information to succeeding generations. They are formed early during the development of the embryo, and eventually divide through meiosis to yield the gametes (sperm or egg).

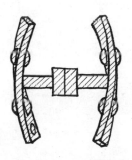

HAPLOID: The genetic content of a single set of chromosomes. Sex cells are haploid and in some organisms (bees and wasps) somatic cells are also haploid. Upon fertilization, the haploid egg receives a second set of chromosomes from the sperm producing a diploid cell. (See DIPLOID).

HISTONES: Proteins (rich in basic, that is, positively charged, amino acids) found in the chromosomes. There are five fundamental histone types. Nucleosomes consist of a helix of DNA wound around a core of histones.

HORMONES: Substances (often small polypeptides or proteins) that are synthesized in one group of cells of the body and then released to affect the functioning of other cell types (or organs) in the body.

HYBRIDIZATION: The formation of double-stranded DNA-DNA, DNA-RNA or RNA-RNA complexes from a mixture of single stranded DNAs or RNAs. Hybrids are formed only if the base sequences are complementary. Also, a technique for determining the sequence similarity between two nucleic acid molecules.

IMMUNOGLOBULINS: Antibodies. They consist of "light" and "heavy" protein chains bound together in a Y-shaped structure.

INDUCERS: Small molecules (often metabolites such as sugars) which bind to repressor protein releasing it from the operator. (See OPERATOR).

INSULIN: A polypeptide hormone secreted by specialized cells in the pancreas that regulates metabolism and the production of energy. It stimulates the uptake of glucose by muscle cells, and the synthesis of protein. Insulin was the first protein to have its complete amino acid sequence determined, a feat accomplished by F. Sanger, in Cambridge.

INTRONS: DNA sequences in eukaryotes that lie within genes, but do not code for protein. In most instances, introns have no apparent function. Their presence "splits" the coding region of the gene into segments called exons. In the synthesis of messenger RNA, introns are copied into RNA, but they are removed by splicing, which restores the continuity of the coding sequence. (See EXON).

215

JUNK DNA: DNA without apparent function. Approximately 90% of animal cell DNA does not code for protein, and is speculated to fall in this class. (See INTRONS and SELFISH DNA).

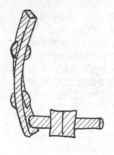

LAMARCKISM: The doctrine, held by Lamarck, that acquired characteristics are inherited. Most biologists consider Lamarckism to be wrong.

LIGASE: An enzyme that joins DNA molecules together. It links the end of one linear double-stranded DNA molecule to the end of another linear double strand to create a continuous double helix.

LIPIDS: Water-insoluble molecules that are important components of cell membranes and serve for energy storage as well. Steroids, fatty acids and waxes are examples of lipids.

LYMPHOCYTES: Cells in the blood which, in the presence of a foreign substance (antigen), will divide and produce antibodies.

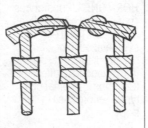

MEIOSIS: The process by which a cell gives rise to daughter cells with half the number of chromosomes. (Diploid cells become haploid.) Sex cells are produced through meiosis.

MESSENGER RNA (mRNA): RNA molecule which is transcribed from a gene, and that contains the coded information for the amino acid sequence of a protein. The information in mRNA is translated on the ribosome. In prokaryotes, one mRNA can encode more than one protein.

MITOSIS: The stage in the life cycle of an eukaryotic cell during which sets of chromosomes destined for daughter cells separate and cell division takes place.

MUTATION: A change in the structure of the genetic material (i.e. in the DNA base sequence), which is inherited. Mutations can alter gene function.

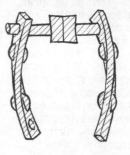

NATURAL SELECTION: In evolutionary theory, the process whereby the adaptation of a population to its environment is improved. A large number of variant forms are produced (through DNA recombination, sexual reproduction, mutation etc.) and those best adapted survive and reproduce, passing on their genetic material.

NONSENSE MUTATION: A mutation that changes a codon into a three base sequence that does not specify any amino acid. Such triplets, known as nonsense codons, are UGA, UAA, UAG.

NUCLEOSOME: The repeating structural unit of

a chromatin consisting of 200 base pairs of DNA wrapped around a histone core. The nucleosomes, plus the DNA that links nucleosomes to one another, make up the chromatin fibers of chromosomes. (See CHROMATIN and CHROMOSOMES).

NUCLEOTIDE: The fundamental unit of the DNA (or RNA) chain. Nucleotides consist of the base (adenine, guanine, cytosine or thymine in DNA, with the latter replaced by uracil in RNA) plus the sugar (deoxyribose in DNA, ribose in RNA) and linked phosphate.

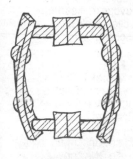

ONCOGENE: Gene responsible for the transformation of normal cells into cancer cells. In some cases, oncogenes are mutant versions of normal cellular genes.

OPERATOR: The site on DNA to which the repressor protein binds, preventing RNA polymerase from transcribing the operon, or structural genes. Thus, a control site for transcription.

OPERON: The groups of adjacent structural genes controlled by an operator and a repressor.

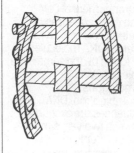

PEPTIDE: A chain of amino acids.

PEPTIDE BOND: A link (convalent bond) between two amino acids in protein.

PHENOTYPE: The characteristics of an organism as manifested in its developed form. The phenotype is the consequence of the interaction of genes with the environment. (See GENOTYPE).

PICOGRAM: One millionth of a millionth of gram.

PLASMID: A small circular DNA molecule (typically about 5000 nucleotides long) that replicates in a bacterium independently of the bacterial chromosome.

POLYMERASE: An enzyme which polymerizes nucleotides into long nucleic acid chains; RNA polymerase synthesized RNA, and DNA polymerase, DNA.

PRIMER: Short DNA or RNA chain, base-paired to a complementary DNA strand, which is elongated by DNA polymerase. The 3' terminus of the primer is the acceptor for the newly added nucleotide residues, and is the start point for DNA synthesis. Reverse transcriptase also uses a RNA primer, but employs RNA as the template.

PROKARYOTES: Organisms whose cells lack nuclei, including bacteria and blue-green algae. (See EUKARYOTES).

PROMOTER: The site on DNA where RNA polymerase binds and initiates transcription. More properly, it is defined genetically as a site whose mutation alters the rate of transcription of an adjacent gene.

PROTEIN: A biological molecule consisting of amino acids linked together in a chain. Proteins range from tens to thousands of amino acids in length, and serve the cell as enzymatic catalysts, structural components, and in the case of peptide hormones, molecules for transmitting informational signals from one part of the body to another.

PSEUDO-GENE: A gene which is non-functional, most often as a result of mutational damage incurred during the course of evolution. Pseudo-genes may arise through duplication of functional genes, followed by divergence of one copy through mutation such that it no longer may be expressed. In some cases, pseudo-genes are formed by copying mRNA into DNA, and inserting the copy back into the chromosomes. These lack introns and have the spliced structure of the mRNA and are called "intron-less pseudo-genes".

PURINES: Organic bases containing both carbon and nitrogen with a two ring structure. Adenine and Guanine are purine bases found in DNA and RNA, linked respectively to deoxyribose and to ribose.

PYRIMIDINES: Organic bases containing both carbon and nitrogen atoms arranged in a single ring structure. Thymine and cytosine are pyrimidines found in DNA linked to deoxyribose. In RNA, the pyrimidine uracil replaces thymine, and like cytosine, is linked to ribose.

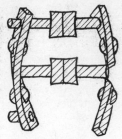

RECOMBINATION: The rearrangement of DNA such that sequences originally present on two different DNA molecules are found on the same molecule. With homologous recombination, DNA sequences are exchanged between two very similar (but not identical) DNAs. With heterologous recombination, the transfer is between DNAs unrelated in nucleotide sequence. Recombination may take place by breakage and reunion of DNA molecules, or by a copy-choice mechanism in which DNA polymerase shifts from one template strand to another in the course of replication.

REGULATORY GENE: Gene that encodes a protein or other factor that regulates the activity of a second gene.

REPLICATION FORK: During DNA replication, the fork is the position on DNA where replication is taking place. The parental DNA strands diverge at this point to serve as template for daughter DNA synthesis, creating a Y shaped form.

REPRESSOR: Protein encoded by a regulatory gene that can either combine with inducer, permitting transcription of structural genes, or bind to the operator blocking RNA polymerase access to the promoter, thereby repressing transcription. (See OPERATOR, PROMOTER).

RESTRICTION ENZYME: Enzymes which cleave DNA at short specific nucleotide sequences. Examples are Eco Ri (for E. coli restriction enzyme I) and Hind III (isolated from the micro-organism Haemophilus influenzae serotype d), which cut respectively at GAATTC and AAGCTT. The very high sequence specificity for restriction enzyme cleavage makes them excellent tools for dissecting DNA.

REVERSE TRANSCRIPTASE: An enzyme contained in certain animal viruses called retroviruses. Starting at a RNA primer, reverse transcriptase will make a DNA copy of an RNA template, a process crucial to the retrovirus life cycle, and useful to the genetic engineer in making DNA clones of mRNA. The flow of information from RNA to DNA is the reverse of the normal information pathway, hence the names reverse transcriptase and retrovirus.

RIBOSOMES: The microparticles in the cytoplasm consisting of RNA and protein, where messenger RNA is translated into protein.

SELFISH DNA: Genes proposed to proliferate within the genome to many hundreds or thousands of copies, but which do not serve a function for the organism. Selfish genes are thus parasites of the genome, and represent the ultimate self-centred biological substance. (See JUNK DNA).

SEXUAL CONJUGATION: In bacteria, the transfer of the bacterial DNA chromosome from a male bacterium to a female bacterium. Male bacteria contain plasmids called fertility factors which mobilize this DNA transfer.

SINGLE-STRANDED DNA (RNA): DNA (or RNA) chain in which the bases are not paired with a second complementary chain. Unlike double-stranded nucleic acid, which forms relatively rigid elongated structures, single strands are floppy and can coil back on themselves easily.

SOMATIC CELL: In multi-cellular organisms, a cell of the soma, or tissues, as opposed to a cell of the germ line. Somatic cells divide and differentiate during development, but under normal circumstances do not exchange genetic information.

STRUCTURAL GENE: A gene coding for a protein, such as an enzyme.

SV40: Simian Virus 40, a virus of monkeys, which also infects tissue cultured cells in the laboratory and has served molecular biologists as a model for gene expression and DNA replication in animal cells. SV40 DNA is a double stranded circle with approximately 5200 base pairs.

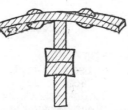

TETRANUCLEOTIDE HYPOTHESIS: A hypothetical structure for the DNA molecule proposed by the organic chemist P.A.T. Levene in which the DNA nucleotides A, G, C and T are arranged in a monotonous repetition of short simple sequences (such as GTACGTACGTAC etc). This hypothesis was incompatible with an informational role for DNA.

THYMINE DIMERS: Two thymine bases on adjacent nucleotides of a DNA strand may be joined together by covalent bonds through the action of ultraviolet light or X-rays to form a thymine dimer. If the dimer is not excised and the DNA is not correctly repaired, a mutational change in the DNA sequence will appear at the site of the dimer.

TRANSCRIPTION: The synthesis of RNA by RNA polymerase, directed by the template strand of DNA. A fundamental step in the utilization of genetic information.

TRANSFER RNA: A class of small RNA molecules, abbreviated tRNA, which function in protein synthesis. tRNAs serve to interpret the genetic code information in messenger RNA.

TRANSFORMATION: The term, as used by Oswald Avery, refers to the transfer of genes in the form of chemically pure DNA to a cell such that they are intergrated into the cell's genome and are functionally expressed. Transformation also refers to the changes in an

219

animal cell's growth properties and morphology which occur when the cell changes from a healthy one to a cancerous cell. This can occur when the cell acquires an oncogene. Cells with unchecked growth are said to be transformed.

TRANSLATION: The reading of the genetic code in messenger RNA during the synthesis of protein. This is performed by tRNA, on the ribosome, and leads to the assembly of amino acids into a chain with amino acid sequence dictated by the DNA sequence of the gene.

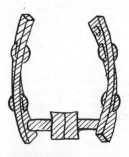

VECTORS: DNA molecules which may be employed to introduce a foreign gene into a cell. Vectors may be designed and constructed by the molecular biologist using genetic engineering techniques. Most are derived from bacterial plasmids, lambda phage, or other genomes which replicate in the cell. They often carry genetic

markers such as a gene conferring resistance to an antibiotic.

VIRUS: Simple microscopic organism consisting of genetic information (most often DNA, but sometimes RNA) wrapped in a protective protein coat. In order to replicate to form new viral progeny, viruses must enter a host cell (animal cell, plant cell or bacterium depending on the virus type) and divert the gene expression machinery (that is, the ribosomes, tRNA etc.) to the manufacture of new viruses. Viruses which replicate in bacteria are often called bacteriophage, or more simply phage.

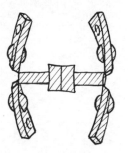

X-RAY DIFFRACTION: When a beam of X-rays passes through a crystal, the beam interacts with the regular array of atoms in the crystal. It exit from the crystal as a complex group of beams. The intensities, angles of exit and phases of these beams are directly determined by the atomic structure of the repeating

unit of the crystal. When changes imparted by the crystal to the beam are determined, the structure of the repeating unit of the crystal may be calculated. This method of structural analysis called X-ray diffraction, was developed at the Cavendish laboratories in Cambridge by William and Lawrence Bragg. The structures of many proteins and of the DNA double helix were determined by X-ray diffraction.

Bibliography

For those who wish to read further about DNA.

Books by and about the scientists, and the history of DNA:
Chargarff, Erwin, **HERACLITEAN FIRE**. Warner Books, New York, 1980.
Jacob, François, **THE LOGIC OF LIFE: A History of Heredity**. Translated from the
    French. Random House, New York, 1976.
Judson, Horace, **THE EIGHTH DAY OF CREATION**. Simon and Schuster, New York,
    1979.
Monod, Jacques, **CHANCE AND NECESSITY**. Translated from the French.
    Random House, New York, 1972.
Sayre, Anne, **ROSALIND FRANKLIN AND DNA**. W. W. Norton, New York, 1978.
Watson, James D. **THE DOUBLE HELIX:** A Norton Critical Edition. Stent, Gunther S.
    ed. W. W. Norton, New York, 1982
Watson, James D. and John Tooze **THE DNA STORY: A Documentary History of
    Gene Cloning**. W. H. Freeman, San Francisco, 1981.

Books with more about the chemistry and biology of DNA:
Stryer, Lubert, **BIOCHEMISTRY**. W. H. Freeman, San Francisco, 1981.
Watson, J. D., **THE MOLECULAR BIOLOGY OF THE GENE**. W. A. Benjamin, Menlo
    Park, 1975.

Illustration Acknowledgements

TO THE MANY WHO HAVE WITTINGLY OR UNWITTINGLY PROVIDED SOURCES OF VISUALS & INSPIRATION:
RENE MAGRITTE, VICTOR MOSCOSO, LAUREN BAN FOON, WALT DISNEY, GIORGIO DE CHIRICO, GUSTAVE DORÉ, LEO MOREY, GRANDVILLE, THE JOLLY GREEN GIANT, MIKE McGUINNESS, MAX ERNST, RICHARD APPIGOLUCKI, THE FRUIT FLY DROSOPHILA, JOHN TENNIEL, THE STATUE OF LIBERTY, TEENAGE PRIEST IN SEX-CHANGE MERCY DASH TO PALACE, MICHAELANGELO BUONARROTI, ANIMAL (GREATEST DRUMMER IN THE WORLD), MC. ESCHER, ESCHERICHA COLI, STANLEY KUBRICK, PAOLO UCCELLO, FRANK R. PAUL, ROBERT CRUMB, AUGUSTE RODIN, "THE WAVENEY CLARION", THE CHIMP THREADING A NEEDLE, WRITERS, READERS (& DRAWERS) EVERYWHERE, AND THE DOUBLE HELIX FOR MAKING IT ALL POSSIBLE. . .

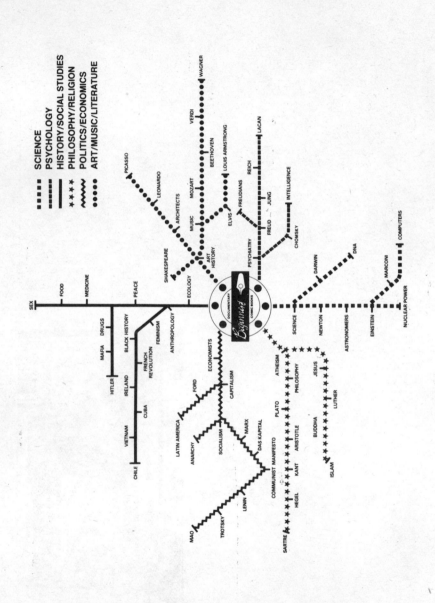